MW01226553

JOURNEY
to the
STARS

Stuart Clark

OXFORD
UNIVERSITY PRESS

OXFORD
UNIVERSITY PRESS

Oxford New York

Auckland Cape Town Dar es Salaam Hong Kong Karachi
Kuala Lumpur Madrid Melbourne Mexico City Nairobi
New Delhi Shanghai Taipei Toronto

With offices in

Argentina Austria Brazil Chile Czech Republic France Greece
Guatemala Hungary Italy Japan Poland Portugal Singapore
South Korea Switzerland Thailand Turkey Ukraine Vietnam

Published by Oxford University Press, Inc.
198 Madison Avenue, New York, NY 10016
www.oup.com

First published 2005

Library of Congress Cataloging-in-Publication Data is available.

ISBN-13: 978-0-19-530515-9
ISBN-10: 0-19-530515-9

1 3 5 7 9 10 8 6 4 2

Printed in Hong Kong

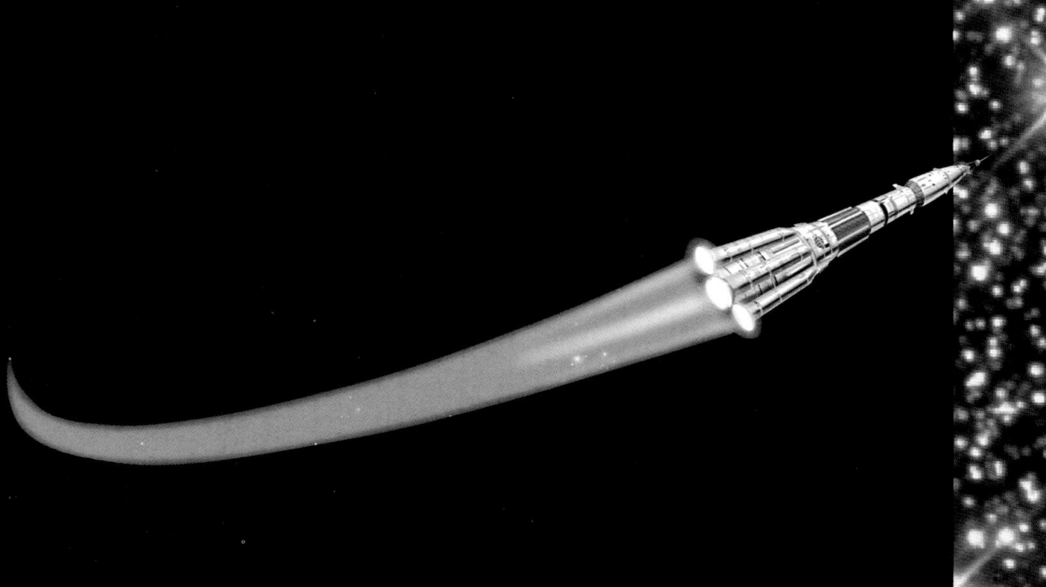

Contents

Astronomical symbols used in this book are explained in the Glossary, pages 46-47

The Night Sky

As night falls, the Sun sets and darkness spreads across the land. Most people retreat into the shelter of their homes but a reward awaits all those who remain outside. On a cloudless night, well away from street lights, the night sky is one of the most beautiful sights a human can see. It is velvet black, dotted with bright stars that shine like diamonds and twinkle softly. In the east, the Moon climbs slowly into the sky. Dark markings on its bright surface make it look like a face, staring down at us from space.

▶ The constellation of Orion, the Hunter. Below the belt stars is a strange fuzzy pink star.

Human eyes adjust to the dark. The longer a person gazes at the night sky, the more they will be able to see. Most people think that all stars are white but looking very carefully brings a wonderful discovery: as your eyes become more sensitive, the stars come alive with colors. Some shine a brilliant blue, others glow yellow, warm orange or gentle red. The blue stars are the hottest stars, yellow stars are cooler and the red stars are the coolest.

▶ Inside the dome is a large telescope, used to study the stars.

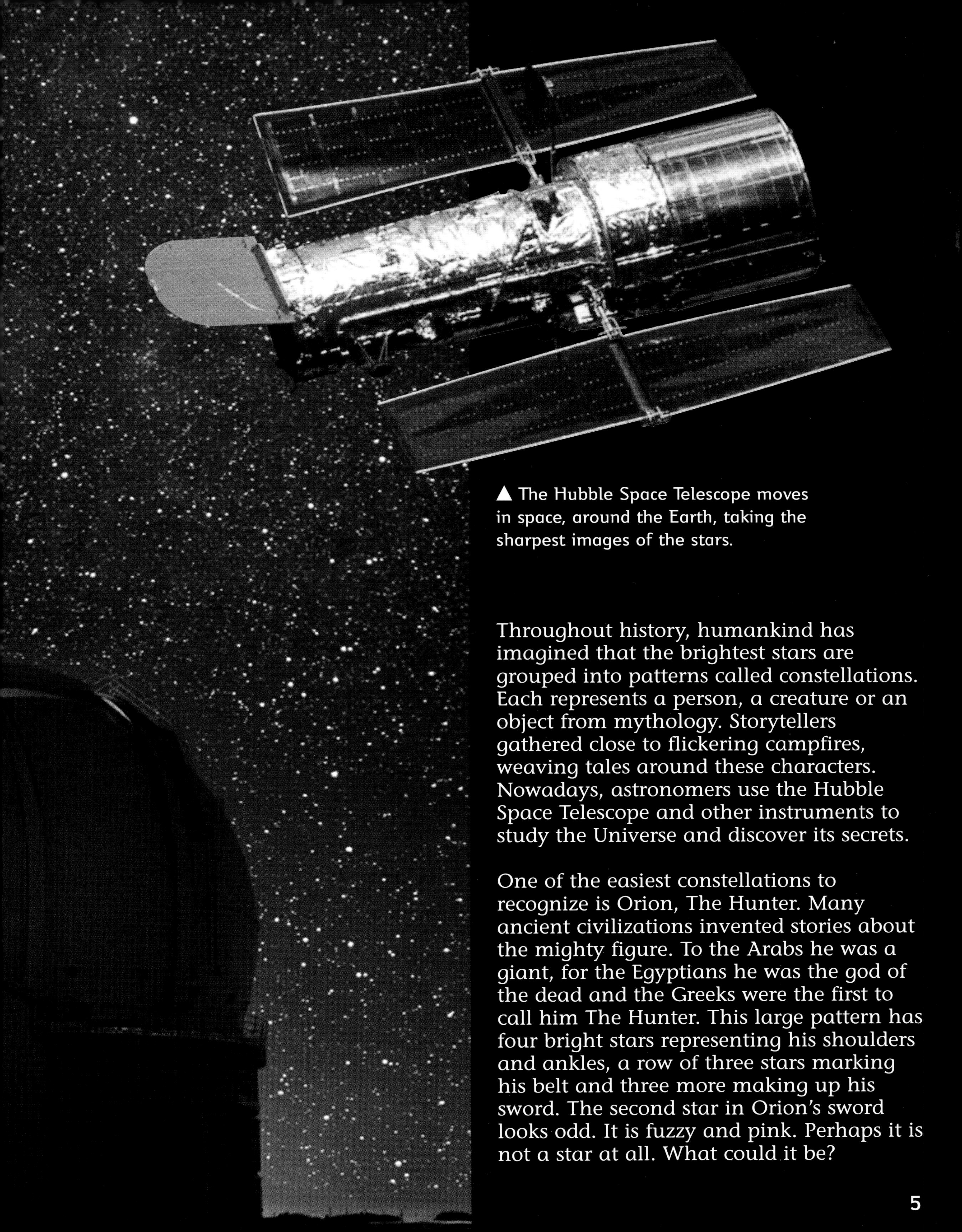

▲ The Hubble Space Telescope moves in space, around the Earth, taking the sharpest images of the stars.

Throughout history, humankind has imagined that the brightest stars are grouped into patterns called constellations. Each represents a person, a creature or an object from mythology. Storytellers gathered close to flickering campfires, weaving tales around these characters. Nowadays, astronomers use the Hubble Space Telescope and other instruments to study the Universe and discover its secrets.

One of the easiest constellations to recognize is Orion, The Hunter. Many ancient civilizations invented stories about the mighty figure. To the Arabs he was a giant, for the Egyptians he was the god of the dead and the Greeks were the first to call him The Hunter. This large pattern has four bright stars representing his shoulders and ankles, a row of three stars marking his belt and three more making up his sword. The second star in Orion's sword looks odd. It is fuzzy and pink. Perhaps it is not a star at all. What could it be?

Map of the Journey

There are many wonderful things in space, such as planets. These are the Earth's brothers and sisters. What are they like? Are they the same or different from our planet, Earth?

To answer these questions, astronomers take close-up pictures of the planets with telescopes and spaceprobes. They use special cameras to see light that is invisible to human eyes. Very powerful computers are then used to examine the pictures to tell us what the planets are really like. Everything in space moves along curved paths called orbits. They are created by gravity. Gravity is a force that tries to pull things together. If objects are moving too fast to be pulled together, gravity sends them into orbit instead.

In our part of space, the Sun has the strongest gravity. The Earth and eight other planets orbit around it. The time taken for Earth to move once around its orbit is known as a year. Together, the Sun and the planets are called the Solar System.

► The journey will take us past the planets of our Solar System and out into deep space. The rocket will pass many stars before leaving them behind to discover what lies beyond.

Imagine flying to the planets in a spacecraft. Then flying on to that strange pink star in Orion. What wonders would be seen? What lies beyond the stars in Orion? More and more stars, perhaps? Does space just stop, or is there even more to be discovered?

◄ Rockets are huge. The human passengers sit at the very top.

We need a rocket to find out. These gigantic machines blast their way into space, breaking free of the Earth's gravity. To live on a rocket, humans must take food to eat, water to drink and air to breathe. The astronauts climb aboard and are strapped in. Banks of computer screens show vital information about the rocket. Everything is set. The countdown begins:

10... 9... 8... 7...

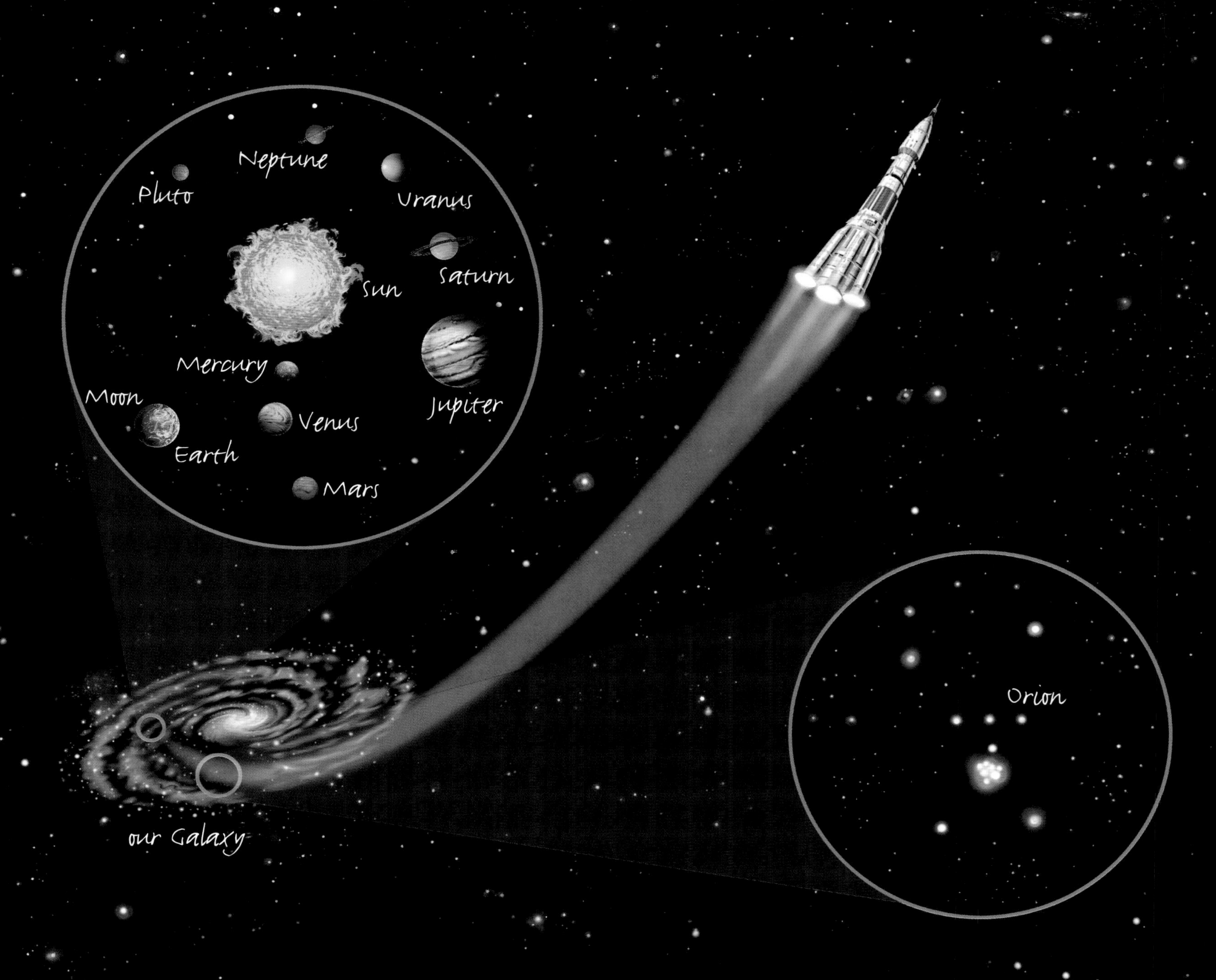

6... 5... 4... 3... 2... 1... LIFT OFF!

Earth

Our Home Planet

The rocket engines burst into fiery life. A deafening roar fills the air. Flames and smoke explode out of the rocket and it begins to push its way into the sky. Higher and higher it climbs, faster and faster it travels. Soon it is traveling eleven miles every second and nothing can stop it now. It breaks free of the Earth's gravity and reaches outer space.

▶ A space shuttle blasts off from Cape Canaveral taking astronauts into space.

From space, looking down, the Earth's oceans are a beautiful, deep blue. The land is a mixture of brown and green. The north and south poles are brilliant white, icy patches. Across this variety of colors, delicate clouds drift through the air. The air around the Earth is called the atmosphere. We can live on the Earth because the atmosphere contains oxygen and nitrogen which humans can breathe.

EARTH

Earth is at just the right temperature for water to flow on its surface. Humans, animals and plants all need water and could not live anywhere without it.

◀ Astronauts in space can watch great storms like this hurricane sweep across Earth.

Sunlight evaporates some of the water in the oceans. That water goes up into the sky and forms clouds. Sometimes these build up into ferocious storms. Thunder crashes and lightning flashes. Rain falls to the ground, quickly running away into the rivers, where it flows out into the oceans and the whole cycle begins again.

If Earth were a lot closer to the Sun, it would be much hotter. More water would turn into steam, and clouds would always cover the sky. The cloudy blanket would stop heat escaping, raising the planet's temperature even higher, until life would become impossible. If the Earth were farther away from the Sun, the temperature would plunge downwards until everything froze, gripping Earth in a terrible ice age. Life would become more inactive before dying out altogether. Luckily, Earth cannot move closer to or further from the Sun.

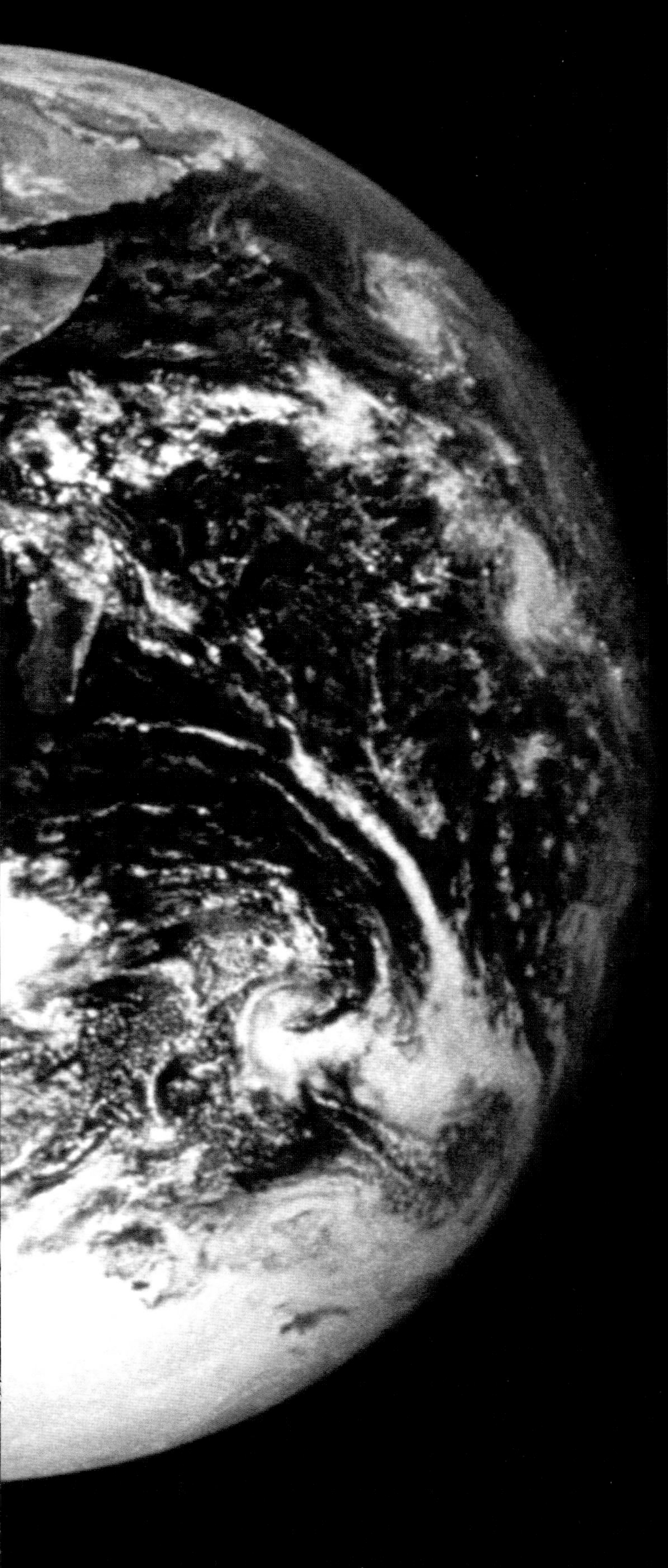

▲ Our world looks beautiful from space. Africa looks brown and Antarctica looks white in this view.

The Moon

▲ The Moon (top) always travels around the Earth.

Orbiting Earth is our nearest celestial neighbor, the Moon. It takes about three days for a rocket to travel there. The Moon is very different from Earth. It is about four times smaller and so its gravity is too weak to attract gases to make an atmosphere. It has no clouds, no oceans and no weather. Swooping down low over the rocks, there is nothing to see but miles of unending desolation. Boulders and rubble lie in the grey lunar dust that covers the whole surface.

A large, flat-topped mountain comes into view, and rises high into the sky. Skimming up the rocky wall, as if on a roller-coaster ride, the summit rushes closer. Flying over the summit, there is nothing but empty space. Instead of having a flat top, the land suddenly plunges down into a large bowl shape, far below. So, this is not a mountain at all but a crater. It was created four thousand million years ago, shortly after the Solar System formed, when a hunk of rock nineteen miles wide hit the Moon. In the gigantic explosion that resulted, huge amounts of Moon rock were blasted upwards. They then fell back onto the surface and became the crater walls.

On the horizon a brilliant glint of reflected sunlight shines for a instant. Something metal is sitting on the surface of the Moon. On closer inspection the remnants of a spacecraft landing are spread across the lunar landscape. An abandoned radio antenna points silently towards space. A forgotten moon buggy is parked close by and the base of a spacecraft rests on the dust. These are the remains of a series of space missions, called Apollo, that landed human beings on the Moon in the late 1960s and early 1970s.

The footprints of those brave American astronauts can still be seen on the surface of the Moon around the landing site. Although they stood on the Moon decades ago there is no wind here to blow their footprints away. They will last for centuries more . . .

◀ This astronaut's footprint will never be blown away on the Moon.

▼ The surface of the Moon is covered with round craters.

► The Apollo astronauts left equipment behind on the Moon.

The Sun

The Sun is the largest object in the Solar System. Imagine putting the Sun onto one side of a gigantic set of weighing scales and all nine planets onto the other. The Sun would still be heavier. In fact, one thousand Earths would have to be added and then one thousand of each of the other planets before the scales balanced.

▼ The Sun is a globe of incredibly hot gas that gives off light and heat.

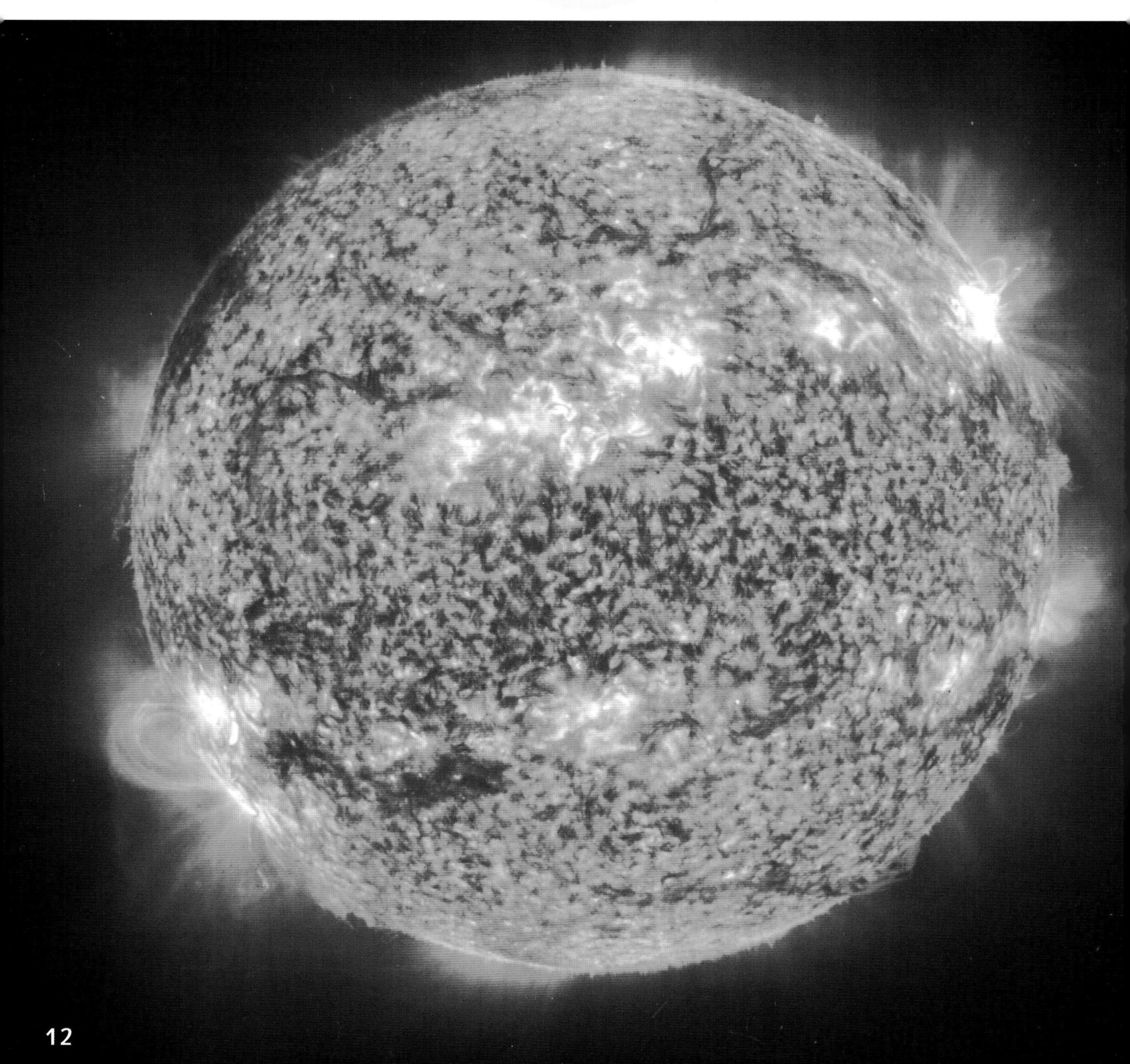

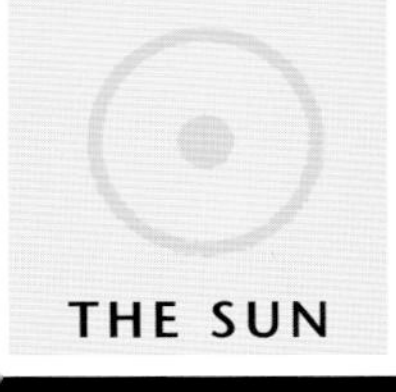

Now imagine flying close to the Sun. This is a dangerous part of space and no human has ever dared venture this close to the Sun. It is a fearsome inferno of heat. But it is not alight, burning like a log fire, instead it is an enormous ball of incredibly hot gas, giving off its energy as heat and light.

The surface of the Sun is over one hundred times hotter than a kitchen oven. Here, the gas behaves like boiling milk, swelling upwards and giving out energy before sinking back down into the interior. Large twisting streamers of hot gas erupt into space, arching round to splash back into the Sun.

Cooler patches on the solar surface, that look dark, are known as sun spots. Without warning a blinding flash of light explodes just above one of these sun spots. It is an erupting solar flare and flings millions of tiny particles into space. These race like a wind through the Solar System, traveling hundreds of miles every second. Any planet that gets in the way is violently bombarded. Particles crash into its atmosphere, creating glowing lights called aurorae.

Having survived this close approach to the Sun, the spacecraft blasts away. It will race towards the planets, then to the stars . . .

IF IT WERE possible to fly into the Sun it would become hotter and hotter until, at the very center, the temperature would reach about ten million degrees centigrade. Here the Sun's energy is generated by tiny particles colliding with one another. This energy tries to escape but, because the Sun is so dense, it takes a million years for it to struggle through the surrounding gas and break out into space.

▼ Our rocket shoots around the Sun and on towards Mercury and Venus.

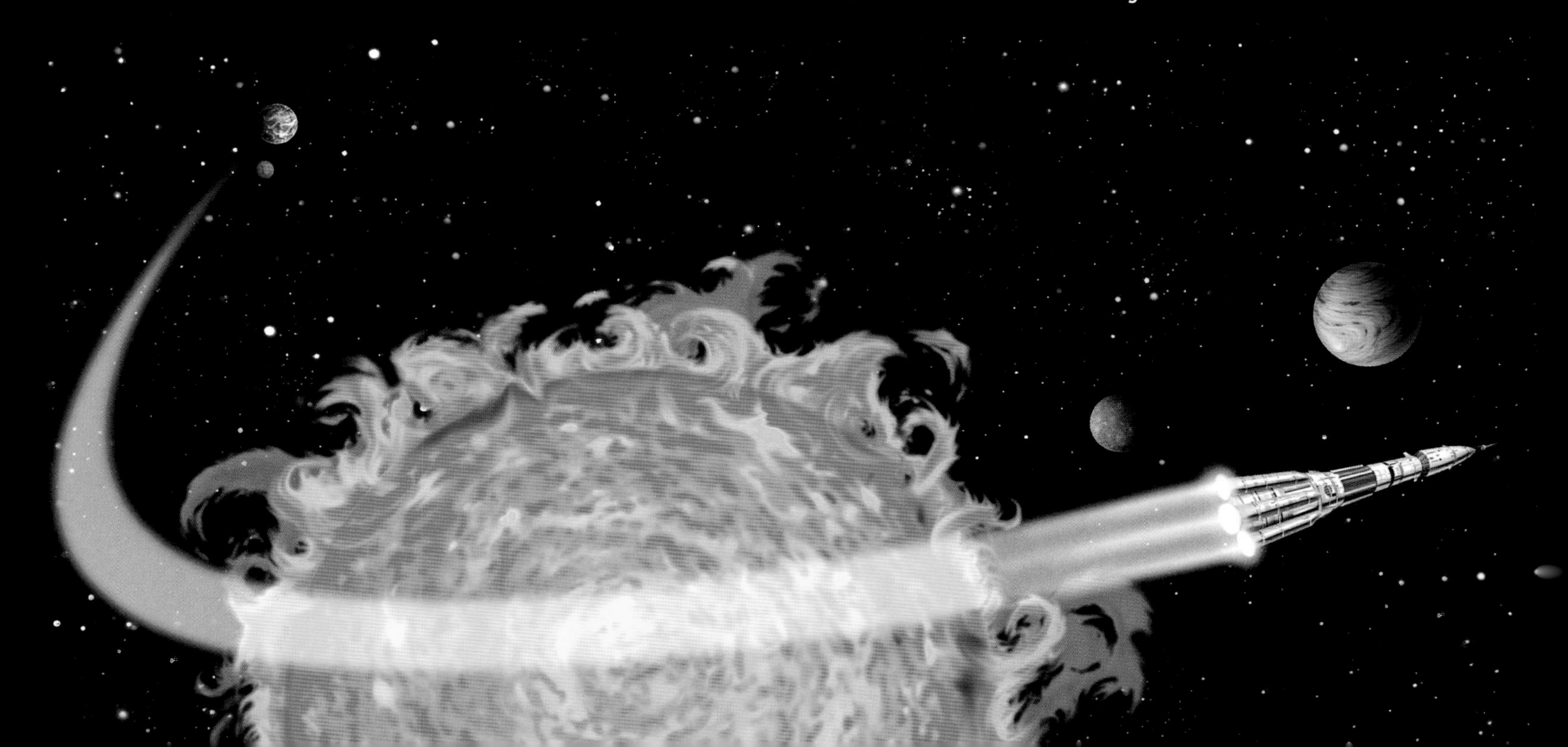

Mercury *and* Venus

MERCURY

The closest planet to the Sun comes into view. It looks very much like the Moon and is about the same size. It is covered in craters and does not have an atmosphere. This is Mercury. The gravity of the Sun holds this world so tightly that Mercury rushes around its orbit in a hectic 88 Earth days. Passing by this planet, the heat of the rocks on the sunlit side changes abruptly to the bitter cold and darkness of the nighttime side of the planet.

▲ Mercury is a tiny planet. This picture is made up of many small pictures stuck together.

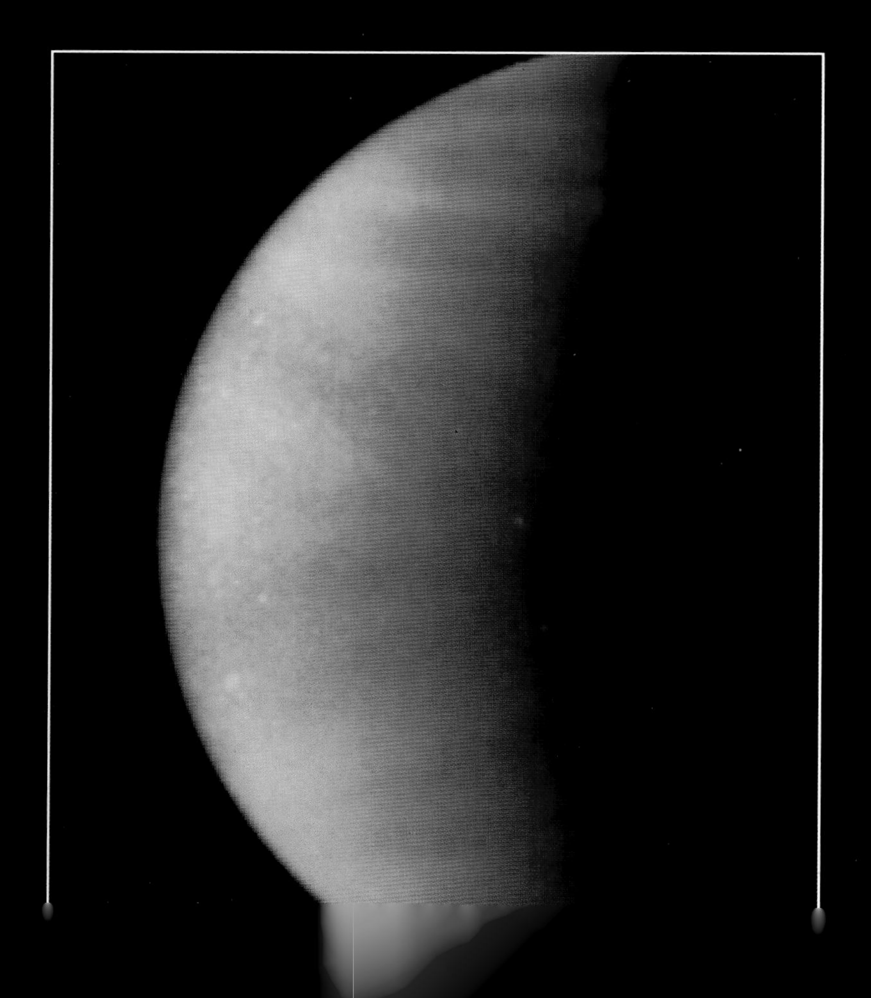

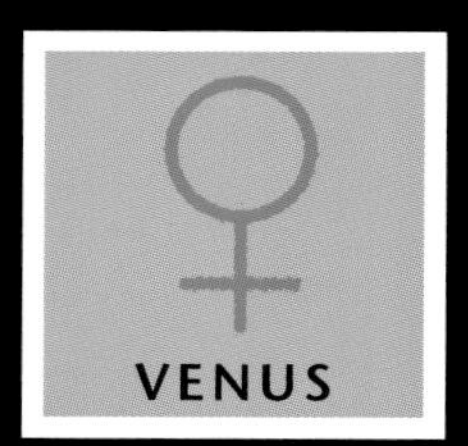

Shooting away from Mercury, the next planet is the mysterious world of Venus. It is twice as far from the Sun as Mercury and much the same size as Earth. Venus is completely covered by swirls of creamy white clouds that wind themselves around the planet. These clouds are so thick that it is impossible to see through them to the surface. Whatever can be down there?

Diving through the cloud tops, a terrifying discovery is made. Venus is a witch's cauldron of nasty chemicals. The clouds contain droplets of sulphuric acid which can combine with another chemical called fluorine and produce one of the most deadly chemicals known to humans. Called fluorosulphuric acid, it is so corrosive that it can eat its way

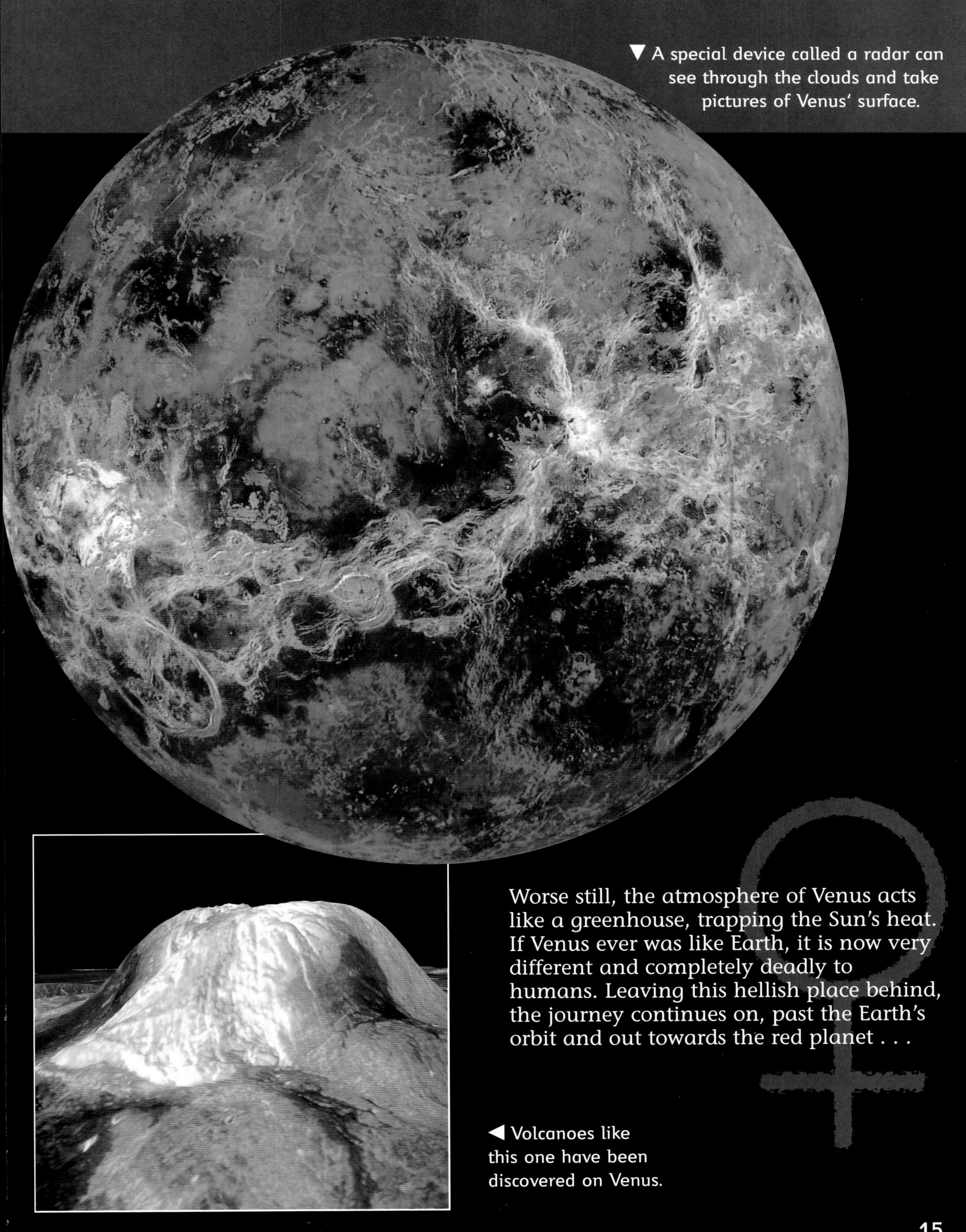

▼ A special device called a radar can see through the clouds and take pictures of Venus' surface.

Worse still, the atmosphere of Venus acts like a greenhouse, trapping the Sun's heat. If Venus ever was like Earth, it is now very different and completely deadly to humans. Leaving this hellish place behind, the journey continues on, past the Earth's orbit and out towards the red planet . . .

◀ Volcanoes like this one have been discovered on Venus.

Mars

The Red Planet

Many astronomers have wondered whether life could exist on Mars. We know now that from space the oceans make Earth look mostly blue but, in the nineteenth century, it was believed that all the plants would make Earth look green. So, in those days, they thought Mars was red because it was covered with red plants!

Approaching Mars, it does indeed look as if it could support life. Like Earth, Mars has an atmosphere and two brilliant white polar ice caps. Huge mountains reach up into the sky and deep valleys cut across the rocky surface. Early morning mist develops in these valleys as the Sun climbs above the Martian horizon.

However, appearances can be deceptive. The planet's atmosphere is too thin and contains the wrong gases, mostly carbon dioxide, for Earth's animal life to breathe. On our planet an outer layer of gas called ozone lets heat and light from the Sun through but blocks out the harmful rays. On Mars there is no ozone layer. There are no oceans, essential to life, and the polar caps contain ice and solid carbon dioxide. Landing on the surface, the mystery deepens: there are old flood plains covered with rocks but still no trace of water. Where could it all have gone?

▲ The Mars Pathfinder spaceprobe was tested on Earth before being sent to Mars.

imagine being able to travel back in time, two billion years. Mars would be very different. It would have a thicker atmosphere, be much warmer and millions of gallons of water would be gushing over the surface. With no ozone layer to protect it, the Sun's harmful rays gradually broke up the water into the gases hydrogen and oxygen. The hydrogen escaped into space and the oxygen sank to the ground, turning the dust and rocks red with rust.

Today, Mars is not only the driest but also the coldest desert that can be imagined. There seems little chance of finding life on this world. Any which may have developed billions of years ago, is almost certainly dead now.

◀ There is a large valley on Mars called the Mariner rift valley. The two spots on the left are extinct volcanoes.

▲ The surface of Mars is covered with rocks and boulders.

The Asteroids

Zooming away from the dead red world, a place is found in the Solar System where no planets have been able to form. At first sight it looks completely empty but, without warning, a huge chunk of rock, the size of a city, tumbles lazily by and gradually fades from view off into the distance. This place is the asteroid belt.

▶ The asteroids are found just beyond the orbit of Mars and stretch to the orbit of Jupiter.

There are thousands of asteroids in the belt but they are very spread out. Isolated and lonely, they wander around the Sun in orbits that are fixed by the gravity of the giant planet Jupiter, which itself orbits the Sun even further out in the Solar System.

Over four and half billion years ago, the asteroids and all the planets were nothing more than tiny bits of dust and gas. They swirled about in a disc, surrounding the young Sun which was forming and heating up in the center. Inside the disc, the tiny dust grains occasionally touched and stuck together. Over the next 50,000 years the dust grains grew into rocky and metallic asteroids.

▲ The asteroid Ida.

▶ The asteroid Gaspra.

The asteroids orbited the Sun, sometimes gently colliding with each other to grow even larger. Eventually they became the planets of the Solar System. Jupiter, Saturn, Uranus, and Neptune were greedy, gobbling up as many asteroids and as much dust and gas as they possibly could, rapidly becoming big, fat planets. Mercury, Venus, Earth, and Mars grew more slowly and, with fewer asteroids to eat, formed into smaller rocky worlds.

The asteroids in the middle, between the two groups of planets, were left behind. Jupiter had grown so large that its gravity threw these asteroids all over the place. Some were thrown out beyond the planets, others smashed into one another at high speed. Any planet that began to form was very quickly shattered back into rubble.

Even today, Jupiter shepherds the surviving asteroids like a flock of sheep, restricting them to certain narrow bands of space. If any move outside these bands, the gravity of Jupiter will toss them away into space.

▲ The asteroid Gaspra tumbles over and over as it moves through space.

The Giant Planet

Jupiter

Beyond the asteroid belt is the outer Solar System. This is the home of four huge planets, the largest of which is gigantic Jupiter. It is so large that more than 1,300 Earths could be crammed inside it. Jupiter is easily recognizable by its bright bands of creamy colored clouds and the darker stripes of clouds between them.

Unlike the Earth, Jupiter has sixteen moons. One of these is called Europa. It is a tiny, frozen ball covered in ice. Below that ice there may be an ocean, like the one under the ice sheets at the Earth's north pole. There may even be tiny creatures, called microbes, living on Europa in the water.

Closer to the titanic planet, another moon, Io, can be seen. On the edge of Io is a small, peculiar bump. It twists and turns as if it were alive.

▼ **Io casts its shadow on the face of Jupiter**

◀ The surface of Europa is covered in sheets of ice.

Astoundingly, it is an active volcano, erupting bright orange lava into space which then falls back down onto the moon and oozes across its surface. The orange lava is made of a chemical called sulphur and there are so many volcanoes on Io that sulphur covers its whole surface. Once on the surface, the sulphur gradually turns darker as the Sun shines on it. There is never much dark sulphur on Io, however, because the volcanoes erupt so often that the older sulphur is quickly covered with fresh, new lava.

Past Io, Jupiter looms closer. The shifting clouds in its atmosphere show just how strongly the tremendous gales, hurricanes and tornadoes blow on this world. The largest of these storms is a monstrous, red whirlwind called the Great Red Spot. It is so large that it could easily swallow up the whole Earth!

No spacecraft could ever land on Jupiter because the planet has no solid surface, just an atmosphere that becomes denser and denser. Anything falling into the planet will sink deeper and deeper into the atmosphere until it is crushed by the weight of the gas pressing down on it.

▲ The gigantic storm, known as the Great Red Spot, on the limb of Jupiter.

▲ Four moons of Jupiter: from the top Io, Europa, Ganymede, and Callisto. None of them is really this close to the planet.

Saturn
The Ringed Planet

Leaving Jupiter behind, another giant world is discovered. The magnificent sweeping curves of Saturn and its rings are a breath-taking sight. It seems much safer than Jupiter. Gliding down into the atmosphere, beautiful yellow clouds of chemicals made from hydrogen gas drift around in the breeze. All seems quiet and peaceful.

Without warning, powerful winds, known as turbulence, suddenly build up. The cloud tops are ripped apart as gleaming white ice crystals are thrown high into the atmosphere. They catch the sunlight and reflect it very brightly all around. The crystals will be blown around by the winds and settle back down deeper into the planet after a few weeks. Storms like this are actually quite rare and many decades pass between them.

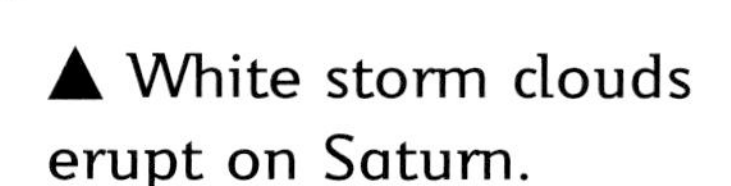

▲ White storm clouds erupt on Saturn.

Pulling up and soaring into deep space again, the rocket passes by the rings of Saturn. They are like a mini asteroid belt made of countless pebbles drifting in orbits high above the equator. Saturn also has a family of eighteen moons. Some of them are tiny, rocky worlds no bigger than asteroids. Others are covered in ice and have had their shiny surfaces shattered by colliding meteorites.

▲ The rings of Saturn are almost flat. When they turn their edges towards our line of sight, they nearly disappear from view.

Saturn's largest moon, Titan, is bigger than the Earth's Moon and is covered in a dense layer of cloud. This makes it difficult to see Titan's surface but, using a special camera, something big that looks like a continent or an ocean can just be glimpsed. It is about the size of Australia.

Titan is an extraordinary place because chemicals containing carbon, another essential ingredient for life, can be found everywhere. They make up the clouds in the atmosphere and drift like fog across the surface. There may even be giant chemical waves crashing along Titan's rocky shorelines. Many scientists believe that this strange, alien landscape is what Earth looked like, four billion years ago, before life began. Unfortunately, Titan is too far away from the Sun and does not receive enough warmth for life to form.

◀ During storms, white ice crystals are thrown high up into the atmosphere.

▶ The mysterious moon Titan will be visited by a spaceprobe in 2004.

Uranus
Neptune *and* Pluto

The cold and lonely outer reaches of the Solar System are the next stage on the journey. Two gas giants, about four times as big as Earth, live in these remote parts. They are called Uranus and Neptune.

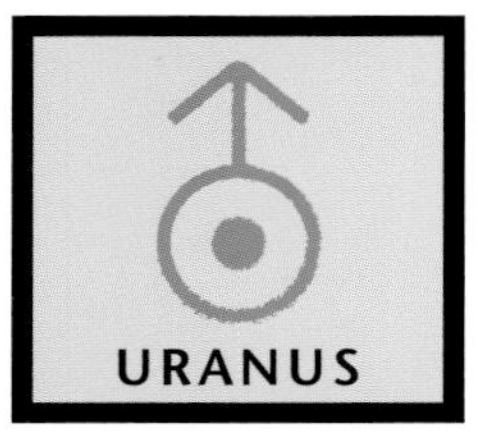

Uranus is the closer one to the Sun and has a smooth, pale blue appearance. Nothing much seems to happen in the atmosphere of this world. At some time in the past, probably when the planets were forming, Uranus was hit by a very large asteroid which caused it to tilt on its side. It now rolls over and over instead of spinning round and round like the other planets.

This means that its equator is top to bottom instead of side to side. The fifteen or so moons of Uranus orbit around its equator so they, too, go over the top of the planet and round underneath.

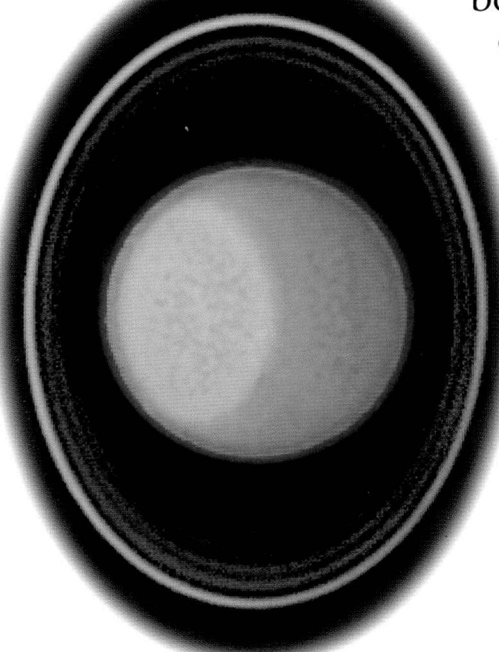

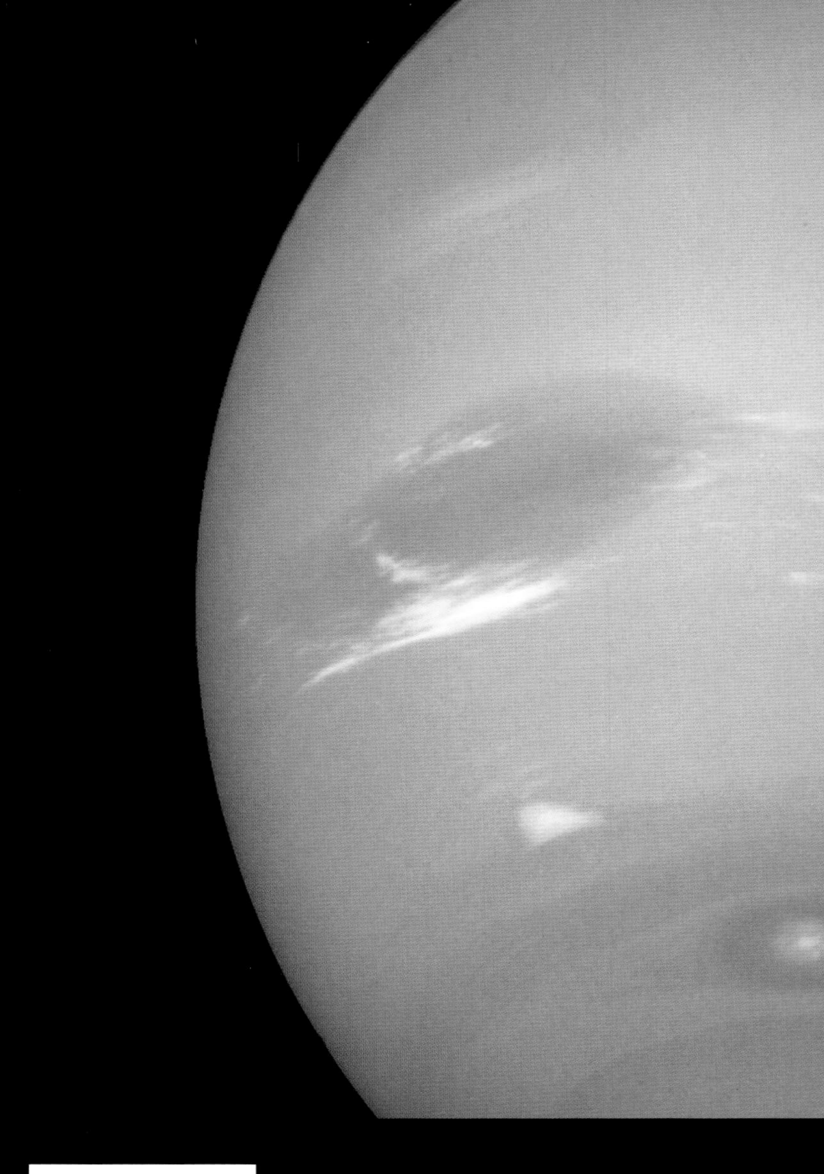

Zooming on, Neptune is much further from the Sun and has an atmosphere that is full of dark cloud belts and swirling tornadoes, rather similar to Jupiter. Neptune is blue like Uranus, its neighbor, but much deeper in color. This is caused by a chemical called methane which turns sunlight blue. High up in this planet's atmosphere, streams of white clouds cast shadows on the blue clouds below.

Neptune's largest and most interesting moon is Triton. Jets of tiny ice fragments erupt near its south pole, leaving dark streaks across Triton's pink surface.

◀ Uranus has rings of pebbles and dust around it. These rings are smaller than those around Saturn.

Even further from the light and warmth of the Sun, tiny, frozen Pluto could be easily overlooked by the unsuspecting space traveler. A rocket would take about forty years to travel from Earth to Pluto. As well as the furthest, it is also the smallest planet in the Solar System. It is so cold that its thin atmosphere of methane is frozen solid to its icy surface.

The journey through the Solar System is almost over. Beyond are the stars and the mystery of that strange, pink star in Orion . . .

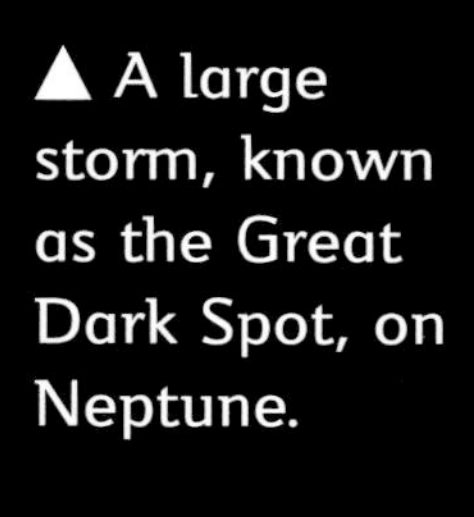

▲ A large storm, known as the Great Dark Spot, on Neptune.

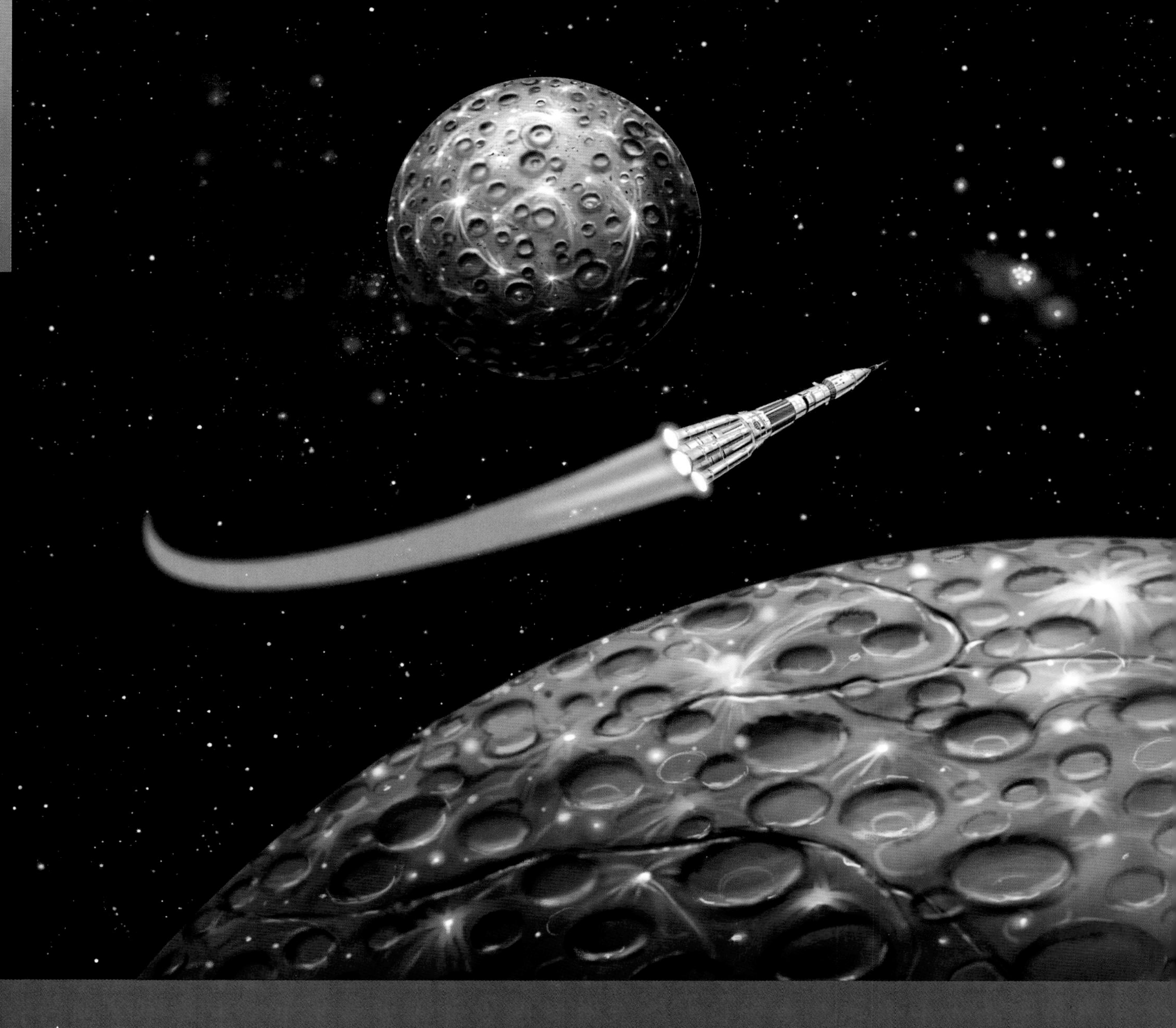

▲ Flying past Pluto (below) and its moon Charon (above), the rocket heads on towards Orion.

Interstellar Space

Just beyond the Solar System, interstellar space begins. This is the almost empty space between the stars. Just ahead, there is a remarkable object. Its shiny, frozen surface is covered in black marks. This is a comet, a gigantic, dirty snowball about six miles wide. It is just one of billions of comets that are spread thinly in a shell, completely surrounding the planets. The shell is called the Oort cloud and is made up of all the asteroids that were thrown away by Jupiter during the formation of the planets.

Most comets take thousands of years to go once around the Sun but a few take less than a century. Sometimes a comet's orbit takes it close to the Sun. For a few weeks the comet will be heated by the Sun's rays and some of the ice will be turned into gas. It will stream away from the comet, making a magnificent tail that can stretch for millions of miles through space. All too soon though, the comet will once again be heading back to the cold space, beyond the Solar System.

▶ A comet throws dust and gas out behind it. The dust curves away and the gas follows a straight line.

Astronomers measure the distance to stars in light years. Light is the fastest thing in the Universe, whizzing around at a speed of 186,000 miles per second. A light year is how far light can travel in a year. It is an amazing 6 trillion miles.

Leaving the last of these comets behind, interstellar space really does become empty. The stars are scattered throughout space, like widely separated bubbles in a vast ocean. The nearest star to the Sun is called Alpha Centauri. It is just over four and a quarter light years away.

Alpha Centauri is a yellow star like the Sun. Unlike the Sun it is not orbited by planets but by two other stars. One is a red star called Beta Centauri and the other is a tiny white dwarf star called Proxima Centauri.

Speeding away from Alpha Centauri, even further into space, it is almost impossible to tell the Sun from the other thousands of stars. The only way not to get lost in space is to look straight at that strange pink star in Orion and head directly for it.

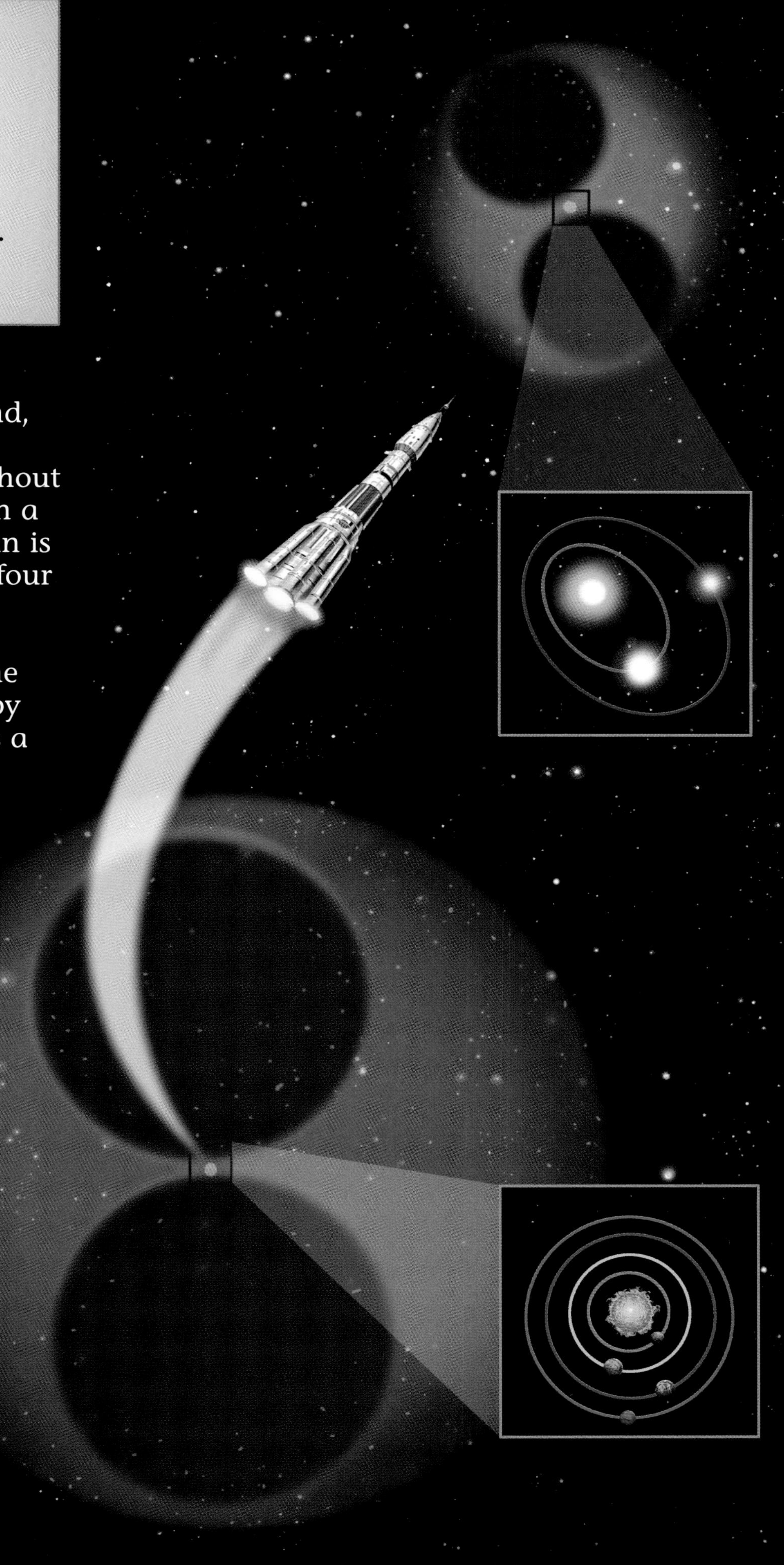

▶ Billions of comets surround the Solar System in a ball-shaped cloud. Other stars are also surrounded by similar comet clouds.

The Birth of Stars

Much further out in space, the strange star in Orion is actually a misty patch of colored light. It is not a star at all, but an enormous cloud of gas millions of times bigger than a star, floating in deep space. It is glowing red, green, and yellow which combine to look pink from the Earth. This is the Orion nebula, which means the Orion cloud. It is made mostly of hydrogen with some other gases. Hydrogen glows red and the other gases glow different colors.

Heading for the heart of the cloud, monstrous fingers of shining gas look as if they are reaching out to pull in passing spacecraft and swallow them whole! Once inside, it is like a gigantic cave with walls of beautiful glowing colors. Many stars live in this cloud. They are packed much closer together than the stars near the Sun.

One of the stars in the Orion nebula glares so brightly that it makes it difficult to see anything else. In fact, the light from this star is so powerful that it is causing the surrounding gas to glow. The star is more than 10,000 times brighter than the Sun and it is young, too. Young for a star, that is. The Sun is about four and a half billion years old but this one is only half a million years old. This is a star nursery!

▶ A dark cloud of dust in Orion makes the shape of a horse's head.

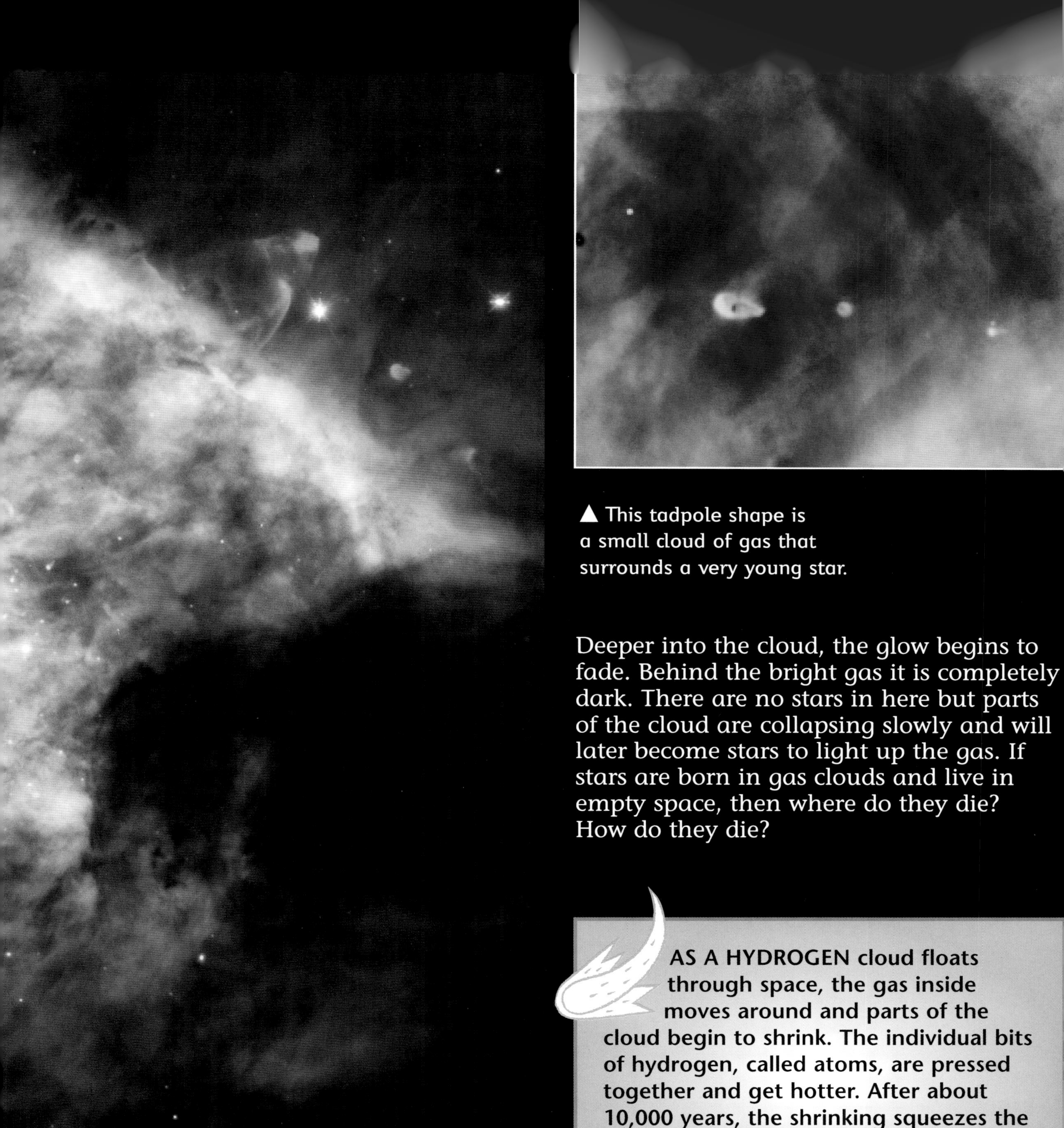

▲ This tadpole shape is a small cloud of gas that surrounds a very young star.

Deeper into the cloud, the glow begins to fade. Behind the bright gas it is completely dark. There are no stars in here but parts of the cloud are collapsing slowly and will later become stars to light up the gas. If stars are born in gas clouds and live in empty space, then where do they die? How do they die?

AS A HYDROGEN cloud floats through space, the gas inside moves around and parts of the cloud begin to shrink. The individual bits of hydrogen, called atoms, are pressed together and get hotter. After about 10,000 years, the shrinking squeezes the atoms so much they stick together. This gives off energy. The star has been born.

▲ The glowing gas of the Orion nebula. This is where new stars are forming.

Dying Stars

Flying out from the Orion nebula, stars appear again. Most lie in a beautiful band that stretches across space. This is called the Milky Way. A vast collection of stars like this is called a galaxy. The Milky Way is our Galaxy.

There are many kinds of stars throughout space, some are similar to the Sun, others are very different. The color of a star shows the temperature of the gas at its surface. A blue giant star is the hottest type of star and a red dwarf is the coolest. Yellow stars, like the Sun, are somewhere in the middle. Counting the number of stars on the voyage shows that hot blue stars are quite rare but there are lots of yellow and red stars.

The hot blue stars blaze energy into space using up all their fuel in less than a billion years. Cooler yellow stars are smaller and shine more gently. They can live for several billion years. Weakest of all are the tiny red dwarf stars which glow only dimly but do so for many billions of years. Blue stars are rare because they only live short lives. But how exactly do any of the stars die?

Flying further through space, it becomes obvious that not all the red stars are dwarfs. Some are much bigger. These big ones are the dying stars. When a yellow or red dwarf star gets old it swells up to become a giant red star. It then splutters out gas and tiny particles into space.

▲ This is the Butterfly nebula, a dying star.

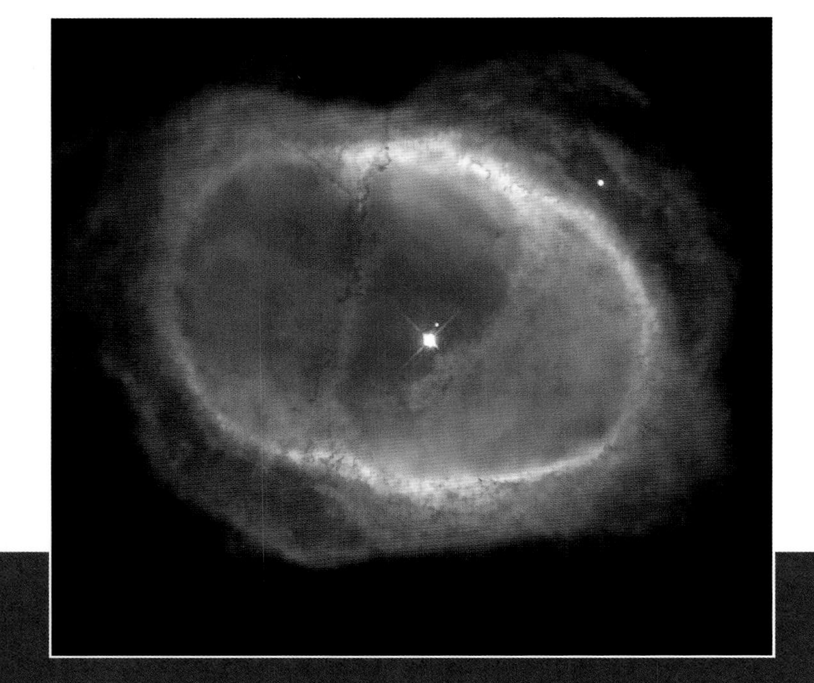

As the star throws off its outer layers, they create delicate patterns of glowing gas. These beautiful displays of color mark the passing away of the star with a reminder of the glorious life it has had. Gradually, over several thousand years, the gas fades and floats away. The small, dead heart of the star is all that remains. Known as a white dwarf, it is a tiny ball of matter that will drift through outer space for evermore.

◀ Astronomers call dying stars "planetary nebulae" because some are round and look a bit like planets. It is a very misleading name.

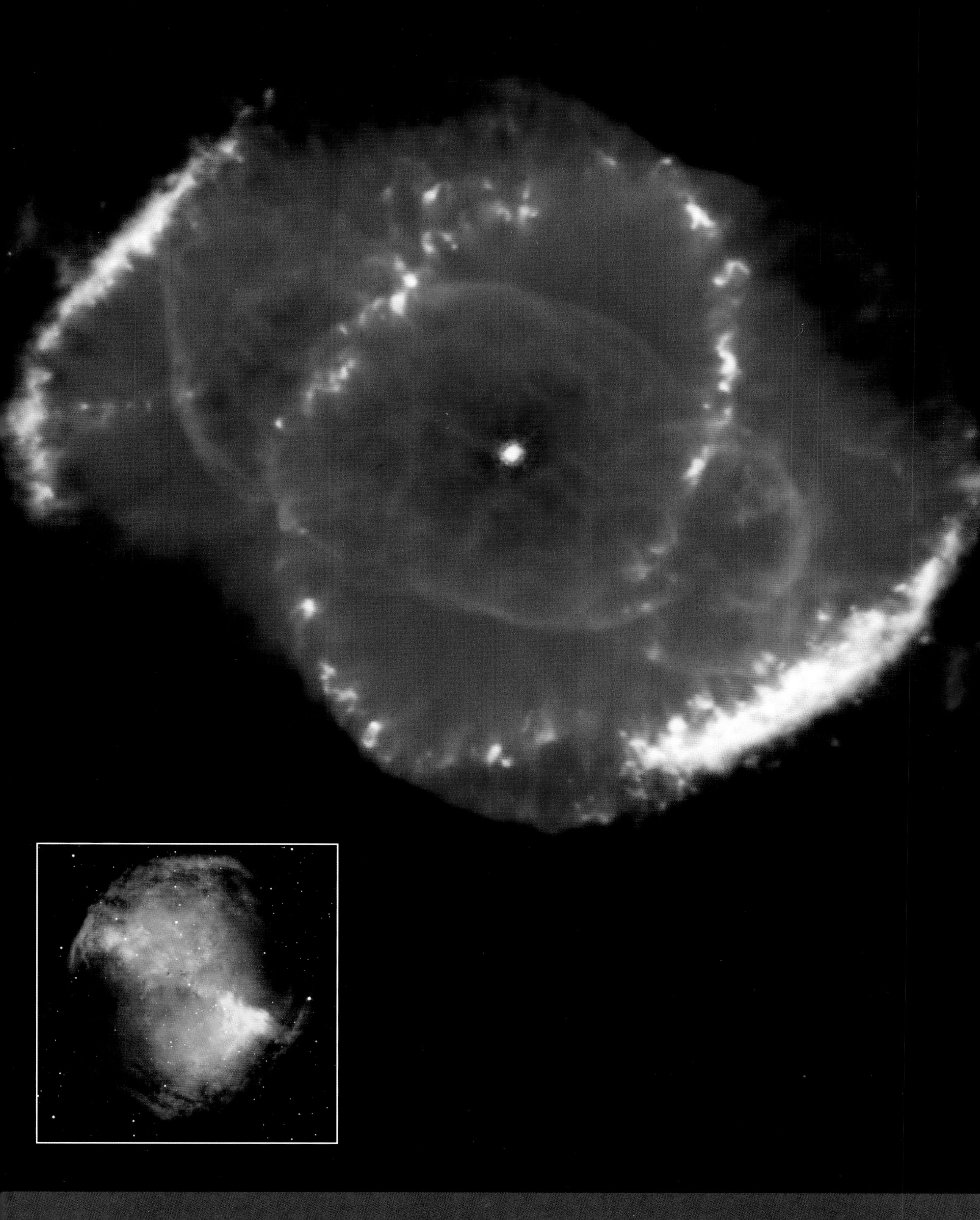

▲ Many different gases glow different colors in the Dumbbell nebula.

▲ The Cat's Eye nebula is a complicated swirl of gases with a white dwarf star in the middle.

Exploding Stars

▼ This is the exploding star Eta Carinae. It could blow itself to bits at any time.

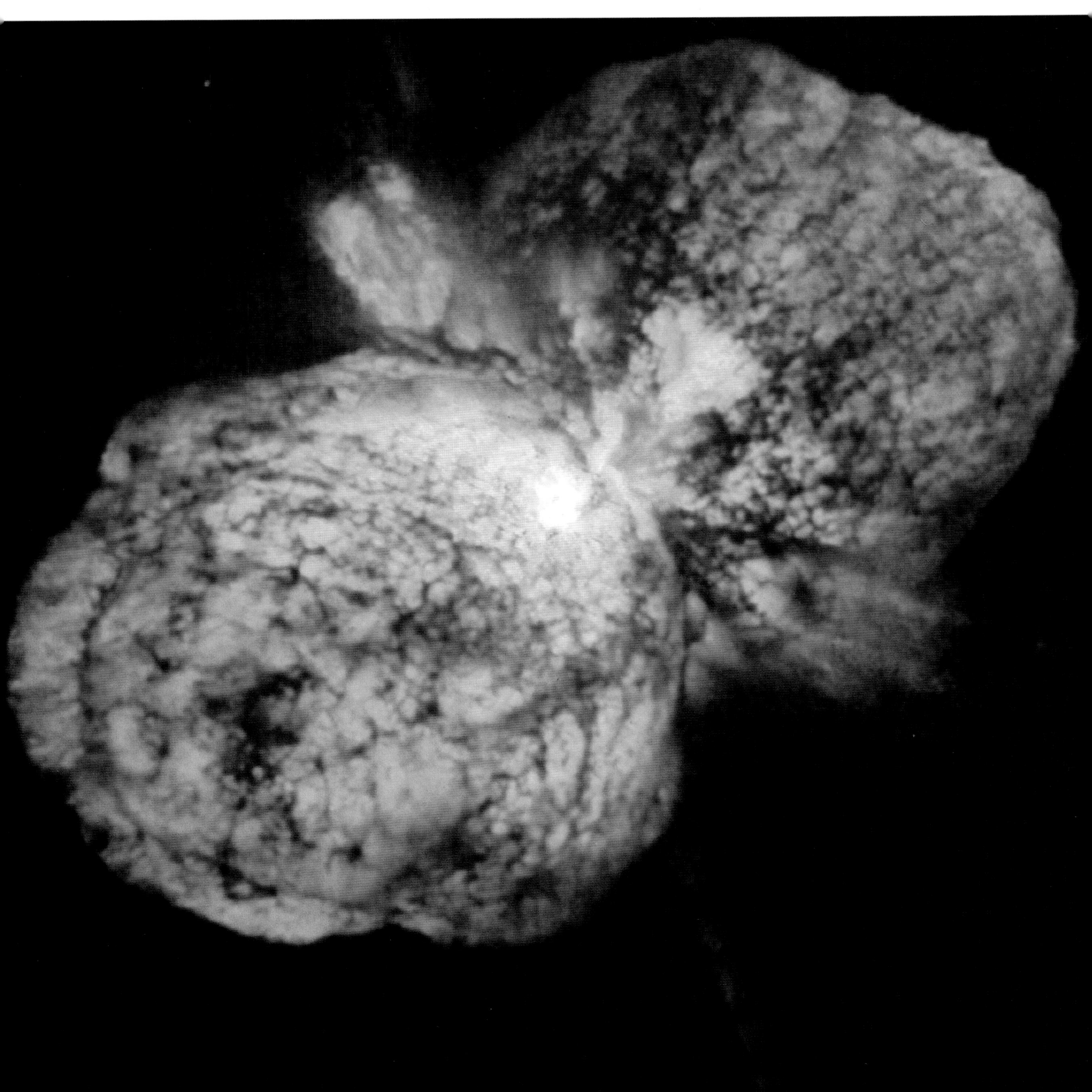

Halfway along the Milky Way a large group of stars bulges out above and below the band of stars. This is the center of the Galaxy, our next destination. Another gas cloud like the Orion nebula is passed. Blue stars can be seen stretching through space in a long curving tail behind the cloud. Following the trail of blue stars, they get older and older, the further from the gas cloud they are found.

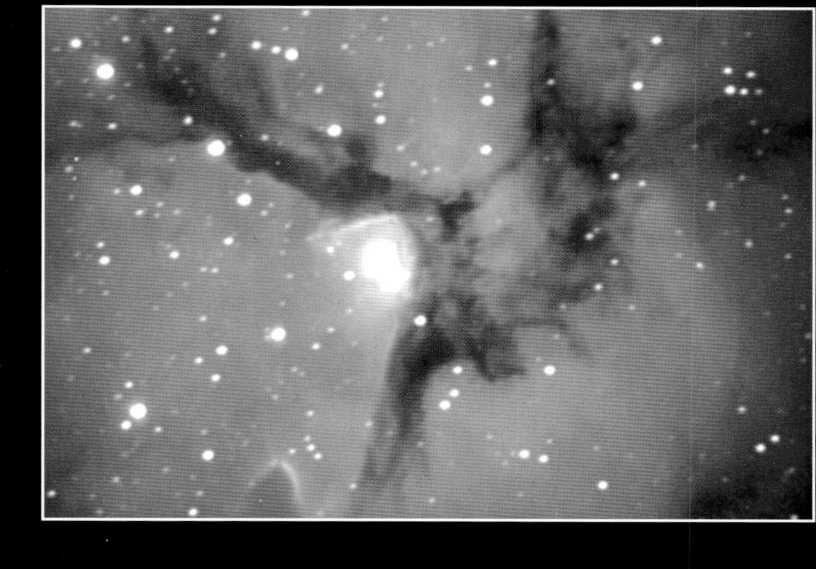

▲ Another glowing cloud of gas with young stars inside.

▲ Blue stars like these will turn red as they grow older and bigger.

Suddenly, a gigantic red star comes into view. It is where a blue star should be. So blue stars must also turn into large red stars as they are dying. This red star is much bigger than any seen so far and it is called a red supergiant.

Flying on, there are fewer and fewer blue stars around. They are all growing old and turning into red supergiants. Near the center of a red supergiant, the star is built in layers, rather like an onion. Different chemicals are found within each of these layers. Right at the center of the star, a core of iron is being made. It builds up in the same way as the ash at the bottom of a fire. Eventually, the iron core grows so large that it can no longer support its own weight. In a tiny fraction of a second it collapses into a tiny ball of matter.

As the iron core collapses, the outer layers of the star race downwards at speeds of up to about 20,000 miles per second. When the gas collides with the collapsed core, an explosion begins that races out through the rest of the star, blowing it to pieces. Such an explosion is called a supernova. A supernova explosion happens about once a century in our Galaxy.

Center of our Galaxy

Flying close to red supergiants on the way to the center of the Galaxy is a very dangerous thing to do. At any moment, one could explode without warning.

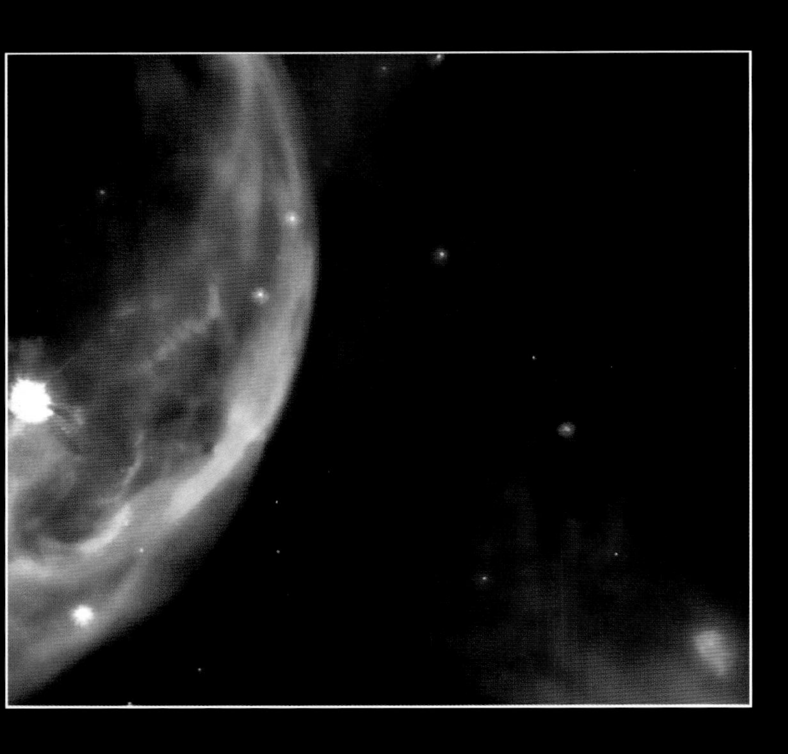

▲ This star is likely to explode at any time.

The surface of a nearby red supergiant begins to bubble in a strange way. In a split second an explosion rips across the star's surface and flings the star's contents into space. As this happens, many different chemicals are made. Anything that gets caught in the supernova explosion will be completely destroyed. For a few days, a supernova gives off as much energy as all the other stars in the Galaxy put together.

A galaxy is rather like a star factory. A few hundred billion different stars are all at various stages of forming, living and dying. When the Milky Way first began making stars, the only chemicals in it were the gases hydrogen and helium. There were no rocks to form planets and no chemicals that could become human beings or other living things. Over billions of years, supernovae have created almost all of the chemicals that make up our bodies. If it were not for these amazing exploding stars, life could never have begun on Earth.

Having escaped the supernova, we can continue the journey to the center of the Galaxy. The river of stars, stretching from the Sun to the center of the Galaxy, is almost 30,000 light years long. Suddenly a brilliant yellow glow of older stars fills the heavens. There are stars everywhere. This is the center of the Galaxy. There are very few blue stars here.

The center of the Galaxy is an enormous cluster of stars. Changing course, our spacecraft blasts away from the Milky Way, out beyond the stars of our Galaxy and into intergalactic space . . .

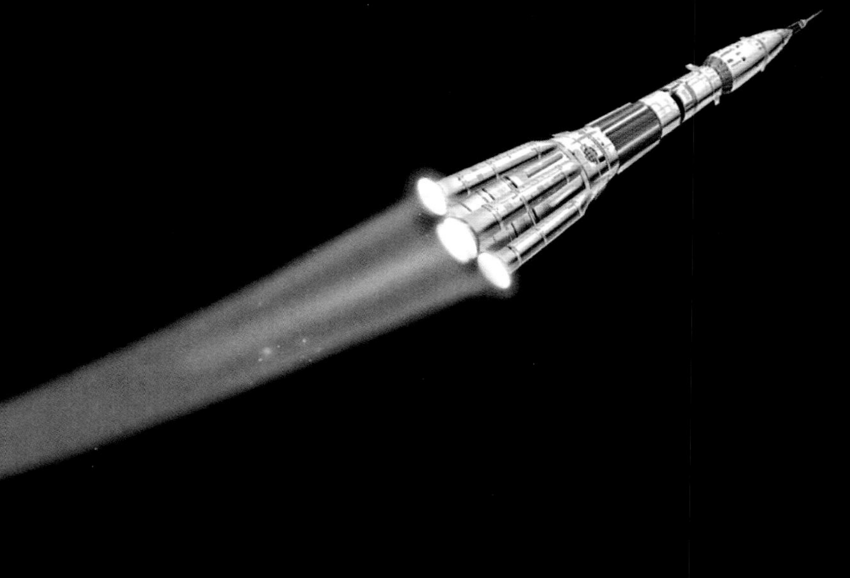

◀ There are many, many stars in the star clouds at the center of our Galaxy.

Intergalactic Space

Zooming upwards and away from the Milky Way, the stars are soon left behind. Having been used to seeing stars all around, the emptiness beyond the Galaxy is frightening. Heading out into intergalactic space, dark objects pass by every now and again. These are the dead remains of the first stars that ever lived.

Looking back, the shape of our Galaxy can be clearly seen. The center looks like a swarm of fireflies frozen in flight. Sweeping out from the edges of the center are large streamers of bright blue stars. They curl around the center of the Galaxy, sweeping outwards in a spiral pattern. Along one side of each spiral arm, the glowing gas of star forming clouds can be seen. The other side of the spiral arm is where the red supergiants turn into supernovae. Between the arms are darker areas where gas clouds form from supernova debris before collapsing to become new stars.

▶ The Earth orbits the Sun in one of the outer spiral arms of our galaxy.

The whole Galaxy rotates slowly. It takes several hundred million years for it to spin around once. During that time, stars are constantly being born while others are dying. The Galaxy is always changing.

So, just what lies out here, beyond the Milky Way?

A cluster of a few hundred thousand stars can just be seen in the distance. There is another, and another, and another. Dozens of clusters in fact. They are all in orbit around the center of our Galaxy and are known as globular clusters. The light from them is yellow, like the center of the Galaxy, showing that there are very few young blue stars in them.

There are a few other faint patches of light out in space. What are these? They do not look like stars. Are they gas clouds? There is one that appears brighter than all the others. It will be the best one to investigate first.

▲ A globular cluster is a collection of hundreds of thousands of old stars.

Other Galaxies

From closer up, the patch of light is not a cloud of gas at all but another collection of hundreds of billions of stars, like the Milky Way. Looking backwards, the now far distant Milky Way itself looks like a faint patch of light. So these faint smudges are the galaxies where stars are huddled together.

This one is the Andromeda galaxy. It is one of the nearest galaxies to the Milky Way. Even so, the distance between them is almost impossible to imagine. More than two million light years of empty space separate the Milky Way from Andromeda. It takes two million years for light, the fastest thing in the universe, to travel between them.

Peering out into deep space again, many other galaxies can be seen. Out here in intergalactic space, there seem to be as many galaxies as there were stars in interstellar space. Looking carefully, it is obvious that not all galaxies are spiral shaped. Some look like globular clusters, only much larger. These are called elliptical galaxies. Yet more galaxies are messy collections of stars and they are known as irregular galaxies.

◀ If we could look at our own Galaxy from very far away, it would look like this one.

There are about twenty galaxies close to the Milky Way and these are called nearby galaxies. They make up a small cluster known as the Local Group. All the other galaxies are much further away but are also found in clusters. The Virgo cluster is 50 million light years away and much bigger than our little group. It contains hundreds of galaxies pulling on each other with their individual forces of gravity.

In our group, even though the Milky Way and the Andromeda galaxy are so far apart, they are still large enough to attract each other with gravity. It will take at least five billion years for them to come together but, when they do, they will suffer one of the most spectacular fates in the Universe. They will collide.

◀ The Andromeda galaxy is the nearest large galaxy to our own.

▲ Gravity pulls galaxies together to form clusters of galaxies.

Colliding Galaxies

When two galaxies collide, they approach each other on a direct collision course. Just as they are about to touch, activity suddenly begins. Stars are flung out behind each galaxy, as if they are growing bright tails. As the galaxies slide into one another, their shapes change and they begin to look more and more like elliptical or irregular galaxies.

Because the individual stars in each galaxy are so far apart, only a very few crash together and become a single, much larger star. Most stars slide by each other, throwing their neighbors into new orbits. The shapes of the colliding galaxies change as the stars begin moving in their new orbits.

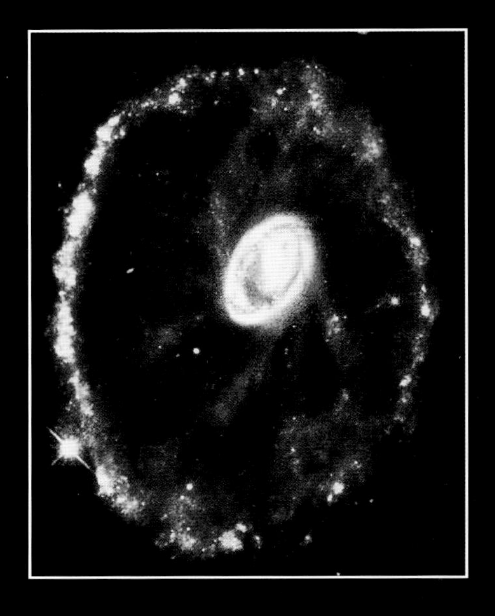

◀ This was once a spiral galaxy but a collision destroyed the spiral arms.

Unlike the stars, the giant clouds of gas in each galaxy cannot avoid hitting each other. In the collision they are suddenly squashed, making them more dense. The denser they become, the easier it is for new stars to form inside them. In colliding galaxies, star formation begins to take place faster and faster. In just a few million years all of the gas in the colliding galaxies turns into stars and the newly joined galaxies blaze splendidly with blue light from the recently created stars. The sudden formation of many more stars than normal is called a starburst.

About twenty million years after the collision, all those massive stars born in the starburst will begin to turn into red supergiants and explode as supernovae. Instead of one supernova every century, as in our Galaxy, a starburst galaxy can have a supernova going off every decade.

After the galaxies have collided and the starburst is over, there will be just one great big elliptical galaxy left, instead of the two separate galaxies.

◀ Two galaxies are colliding. They have formed tails made of stars.

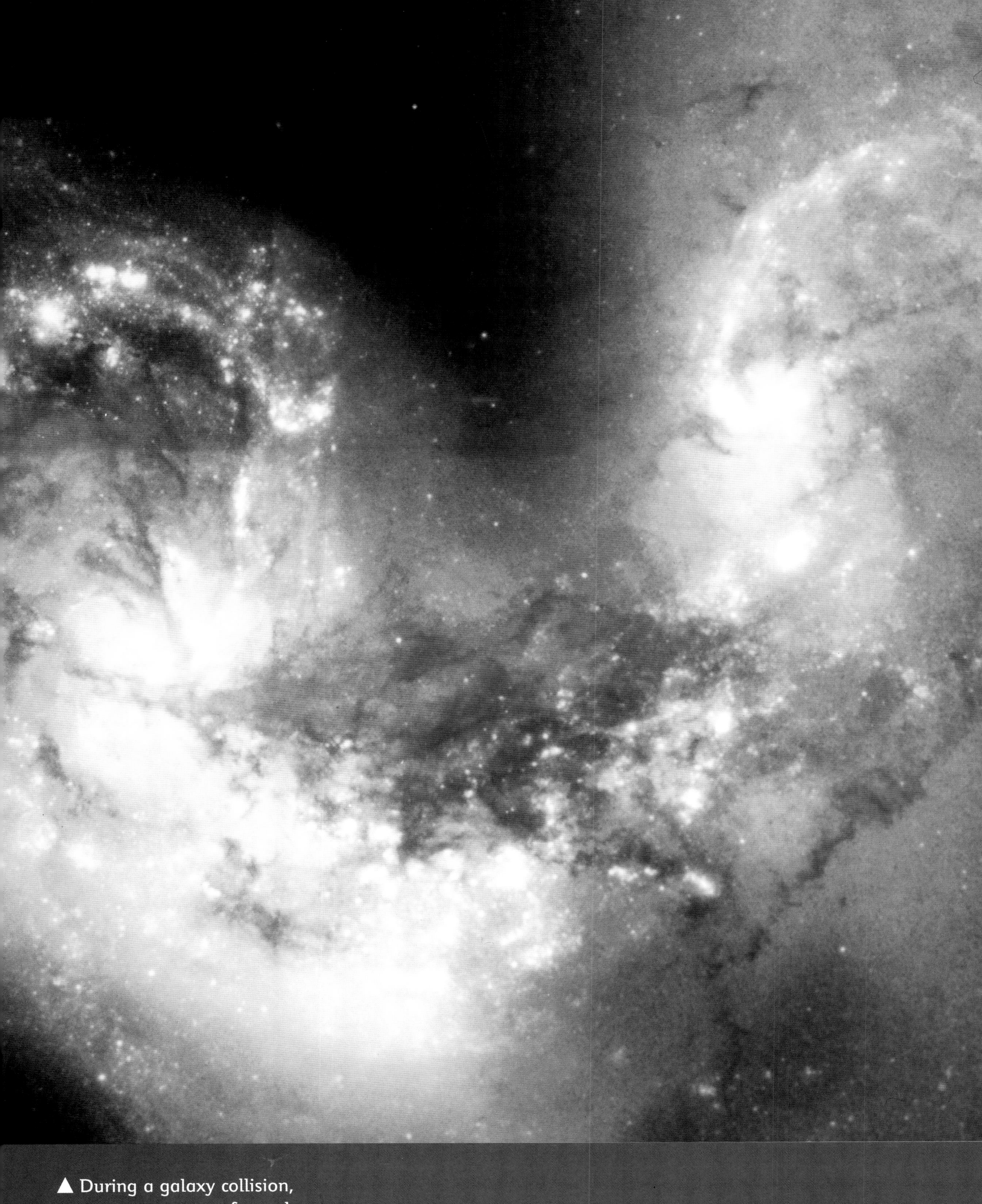

▲ During a galaxy collision, many new stars are formed. This is called a starburst.

Black Holes

It would be almost impossible to travel any further into the Universe. The distances between the clusters of galaxies are just too big. Instead of making the journey to other places, telescopes such as the Hubble will have to be used to look at them. Some galaxies in the Universe are very puzzling. Instead of giving out just starlight, each of these particular galaxies blazes x-rays into space from a tiny spot at their center. They are called active galaxies.

If it were possible to travel into the center of one of these galaxies, it would be very frightening indeed. At the heart of every active galaxy lurks a very strange object: a black hole.

A black hole is an invisible object with such strong gravity that it can destroy absolutely anything that strays too close. One can be formed when a red supergiant star becomes a supernova. Just before the star explodes, the iron core collapses into a tiny ball of dense matter. Sometimes it is so dense that its gravity becomes incredibly strong and it turns into a black hole. A black hole will usually be a few times heavier than the Sun but squeezed into a tiny ball, just a few miles across.

► Gas is sucked into a black hole in a cosmic whirlpool.

Anything could be made dense enough to become a black hole. To make the Earth into one, it would have to be squeezed until it was the size of a small coin. In the center of an active galaxy, a black hole would be very much bigger. It can be made of matter from more than one billion stars and will be about the size of our entire Solar System.

Any stars and gas clouds that pass too closely will be grabbed by the black hole's gravity and ripped to pieces as they are thrown into a whirlpool of hot gas, spiraling around the black hole. The gas becomes so hot that x-rays are given off into space. They are so powerful that they can easily be seen half way across the Universe.

Unfortunately, nothing can save the gas from total destruction. It disappears down into the black hole, like water disappearing down a gigantic cosmic plug hole, and is gone forever.

▲ The center of an active galaxy glows brilliantly with light and x-rays.

Exploring Deep Space

From Earth, only the stars in our own Galaxy can be seen with the naked eye. But we now know that beyond the stars lie hundreds of billions of galaxies. Each of these galaxies contains a few hundred billion stars.

Looking deep into the Universe is like looking far back in time and seeing how galaxies appeared millions and billions of years ago. This is because the distances are so large that even light takes a very long time to travel from them. Telescopes can see billions of light years into the Universe and so they can see galaxies as they appeared billions of years ago.

It seems as if the Universe itself was born more than ten billion years ago. A few billion years after that, there were still no grown up galaxies. In those days, the galaxies were mostly small, young objects. Some got bigger by feeding from the dust and gas in space around them. They gradually formed new stars and turned into graceful spiral galaxies like our own Milky Way.

Others charged around colliding with other young galaxies. These ones grew up fast because they formed stars very quickly in starbursts. Instead of turning into the beautiful spirals, they became big, fat elliptical galaxies.

▲ The New Technology Telescope in Chile tests the latest instruments for studying the stars.

So, what lies beyond the galaxies? As yet, no one really knows but a great number of astronomers all over the world are working very hard to find out. Some of them are building spaceprobes and some are building ever more powerful telescopes to look further and further, to discover more and more. The Very Large Telescope is the biggest telescope system in the world. It is a collection of four giant telescopes that all work together to see incredibly faint objects in great detail.

One day humans may even be able to build starships like the ones in stories and really travel to the stars, to places that, for the moment, we can only gaze at in wonder.

◀ The Gemini Telescope on Hawaii will study deep space, looking at galaxies containing millions of stars.

Glossary

active galaxy A galaxy that sends out x-rays as well as starlight.

Alpha Centauri The nearest star to the Sun.

Andromeda galaxy The nearest large galaxy to the Milky Way.

asteroid A large rocky body, not big enough to be called a planet.

asteroid belt A region of space where hundreds of thousands of asteroids orbit the Sun.

astronaut A person who flies beyond the Earth's atmosphere into space in a rocket. Some of them work on space stations.

atmosphere Gases trapped around a planet. Earth's atmosphere contains gases that humans can breathe. Atmospheres around other planets contain gases poisonous to humans.

atoms The smallest chemical parts of anything.

billion One thousand million.

black hole An invisible object with immensely strong gravity. A black hole is powerful enough to suck in and destroy anything near to it.

blue stars The hottest stars with the shortest lives.

carbon A basic chemical necessary for life on Earth.

carbon dioxide A gas made from one carbon atom and two oxygen atoms. Humans breathe out carbon dioxide.

colliding galaxies Two galaxies that are colliding with each other.

comet A small icy, rocky body in a very large orbit around a star.

constellations Groups of stars, given names by ancient humans who could see patterns in them.

crater A big hole, caused when asteroids or comets crash into planets.

decade Ten years.

desert A bleak place with no water where nothing can live.

disc A flat round shape.

Earth Our home planet. It is the third from the Sun in a system of nine.

elliptical galaxy A collection of hundreds of billions of old stars. An elliptical galaxy can be shaped like a football, or like a soccer ball.

energy Waves given out as heat or light by a body.

equator The imaginary ring around the fattest, middle part of a planet.

flood plains Flat areas of land that have been flooded in the past.

fluorine A highly poisonous gas.

fluorosulphuric acid
The deadliest liquid known to man. It can destroy anything and is found on Venus.

galaxy A group of stars attracted to each other by gravity. Galaxies are elliptical, spiral or irregular.

gas A substance like the air.

gas giant A large planet made mostly of gas.

globular cluster An individual group of stars that orbits a galaxy.

gravity The pull one object has on another. The denser an object is, the stronger its pull of gravity.

Great Red Spot A colossal storm on the planet Jupiter. Humans have been observing it through telescopes for hundreds of years.

Hubble Space Telescope A very advanced telescope that "sees" clearly because it is outside the atmosphere of the Earth.

hydrogen The lightest atom in the Universe.

intergalactic space The spaces in the Universe between each individual galaxy.

interstellar space The spaces in a galaxy between each individual star.

iron A heavy metal.

irregular galaxy A galaxy that is neither elliptical nor spiral in shape.

Jupiter The biggest planet in our Solar System. It is known as a gas giant and is the fifth planet from the Sun.

lava Molten rock, which escapes from underneath the crust of a planet through a hole known as a volcano.

light year A light year is a distance not a time. It is the distance traveled by light in one year. Light travels very, very fast, 186,282 miles every second.

Local Group The nearest galaxies to our own Milky Way. They are all attracted to each other by the force of gravity.

lunar Anything to do with the Moon.

Mars The red planet, fourth from the Sun in the nine planets that make up our Solar System.

The symbols that appear in this book are used by astronomers to stand for the bodies in our Solar System.

Earth Showing the four corners of the Earth.

Moon Shown as a crescent moon.

Sun The center of the Solar System.

Mercury Named after an ancient god, the symbol represents Mercury's staff

Venus Named after an ancient goddess, the symbol shows a mirror.

Mercury The nearest planet to our Sun, it goes once round the Sun in just 88 days.
meteorite A small body that has travelled through space and crashed onto a planet or moon.

methane A gas often found in its icy form in the outer regions of the Solar System.

microbe A tiny, simple form of life.

Milky Way Our own group of stars, sometimes known as the Galaxy. It is spiral in shape.

million One thousand thousands, or ten times ten times ten times ten times ten times ten!

Moon The rocky body that orbits our Earth. Other planets have moons, too. Some of them have lots of moons but Earth just has one.

Neptune The eighth planet from the Sun. It has lots of methane in its atmosphere so it looks dark blue.

Oort cloud The shell of comets out beyond the planets in the Solar System.

orbit The path an object takes through space. If the gravity of one body is not strong enough to make a smaller body crash into it, then the smaller body will travel round and round the larger body.

Orion A large constellation, thought to look like a giant figure.

Orion nebula A large gas cloud in space where stars are forming.

oxygen A gas that humans breathe to stay alive.

ozone layer A layer around the atmosphere of the Earth which lets through warmth and light but prevents harmful rays from reaching the surface.

planet A round rocky body, often with an atmosphere, which orbits a star.

Pluto A very small rocky planet, ninth in order from the Sun.
polar caps At the top and bottom of a planet it is usually very cold so any gas or liquid becomes solid. On Earth the polar caps are made of iced water, on Mars they are iced carbon dioxide.

red dwarf star A tiny star that will live for many billions of years.

red giant star A large star that will become a white dwarf star.

red supergiant star A large star that will become a supernova.

rocket A craft, driven by very powerful engines, which can escape from the Earth's gravity and travel out into space.

Saturn The sixth planet from the Sun in our Solar System. It is known as a gas giant and has a wonderful ring system which can be seen through a telescope.

solar Anything to do with the Sun.

solar flare A sudden explosion on the surface of the Sun that throws hot gas into space.

Solar System The Sun, its nine planets, their moons, the asteroids and comets.

space The emptiness between objects in the Universe.

spaceprobe A robot sent into space to gather data about other objects.

spiral arm A curving band of blue stars which stretches out from the centre of a spiral galaxy.

spiral galaxy
A galaxy shaped so that it has a central bulge with arms of stars curving outwards from the center.

star A very hot ball of gas that generates energy in its center.

starburst
The rapid formation of lots of stars together.

sulphur A chemical often given out by volcanoes. It is pale yellow in color.

sulphuric acid A liquid formed from sulphur, oxygen, and hydrogen.

Sun Our own star, which the planets, asteroids and comets all orbit.

Sun spots Cooler areas of the Sun's surface which look like dark patches.

supernova The catastrophic explosion of a huge old star.

telescope An instrument made of lenses and mirrors which helps humans to see far into space.

temperature A measure of how hot or cold an object is.

Titan A moon orbiting Saturn. It might look like Earth did four billion years ago.

Triton Neptune's largest moon.

Universe Our word for all the galaxies and space that exists.

Uranus The seventh planet from the Sun. It spins over and over instead of round and round.

Venus The second planet from the Sun. Its complete blanket of cloud has made it very hot indeed.

Very Large Telescope A collection of four telescopes in Chile.

volcano A hole in the crust of a planet or moon through which molten rock from beneath can escape.

water The liquid that fills Earth's oceans and seas. It is also found on some other planets and moons, mostly as ice.

white dwarf star The remains of a dead star.

x-ray A very powerful wave of energy.

yellow stars Old stars of medium temperature. The Sun is a yellow star.

♂ **Mars** Named after the god of war, the symbol shows a shield and spearhead.

♃ **Jupiter** Named after an ancient god, the symbol represents a thunderbolt.

♄ **Saturn** Named after an ancient god, the symbol represents a scythe.

♅ **Uranus** A made-up symbol to fit with the others representing planets.

♆ **Neptune** Neptune was god of the seas, the symbol shows his trident (a huge fork).

♇ **Pluto** P+L, the initials of Percival Lowell who searched for the ninth planet.

Index

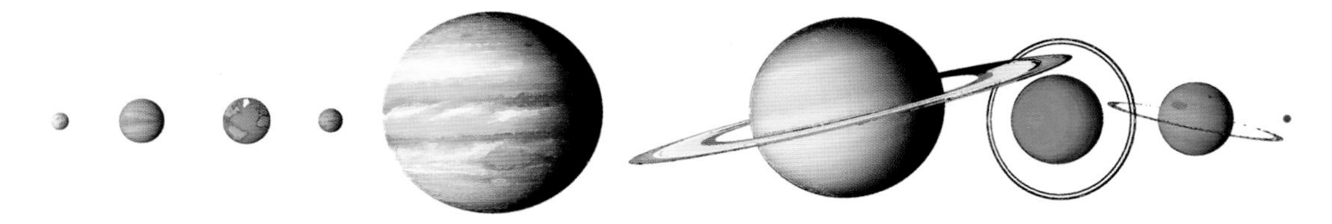

Picture Acknowledgments

All pictures are courtesy of NASA/STSci except:
pages 4 & 5(c), 28(b), 33 & 37 Nik Szymanek & Ian King; 10(b) Telegraph Colour Library; 30(t), 31(bl), 39(t) & 45 courtesy of ESO; 12 courtesy of ESA; 44(b) courtesy of Gemini Observatory; 38 courtesy of AURA/NOAO 39(bl) courtesy of NAOJ.

Artwork by Bob Corley, Clive Goodyer & Julian Baker.
Designed by John Christopher of WHITE DESIGN.

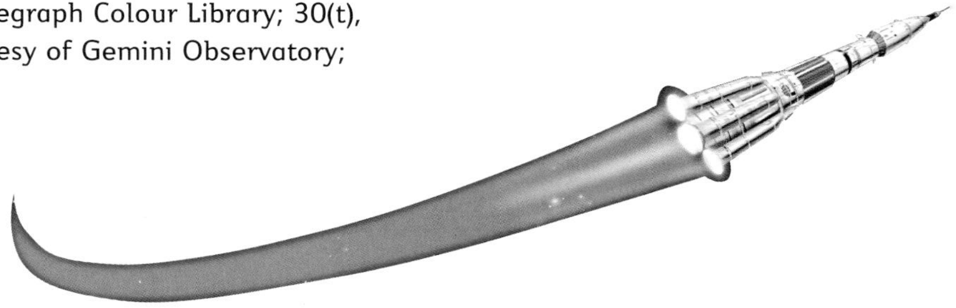

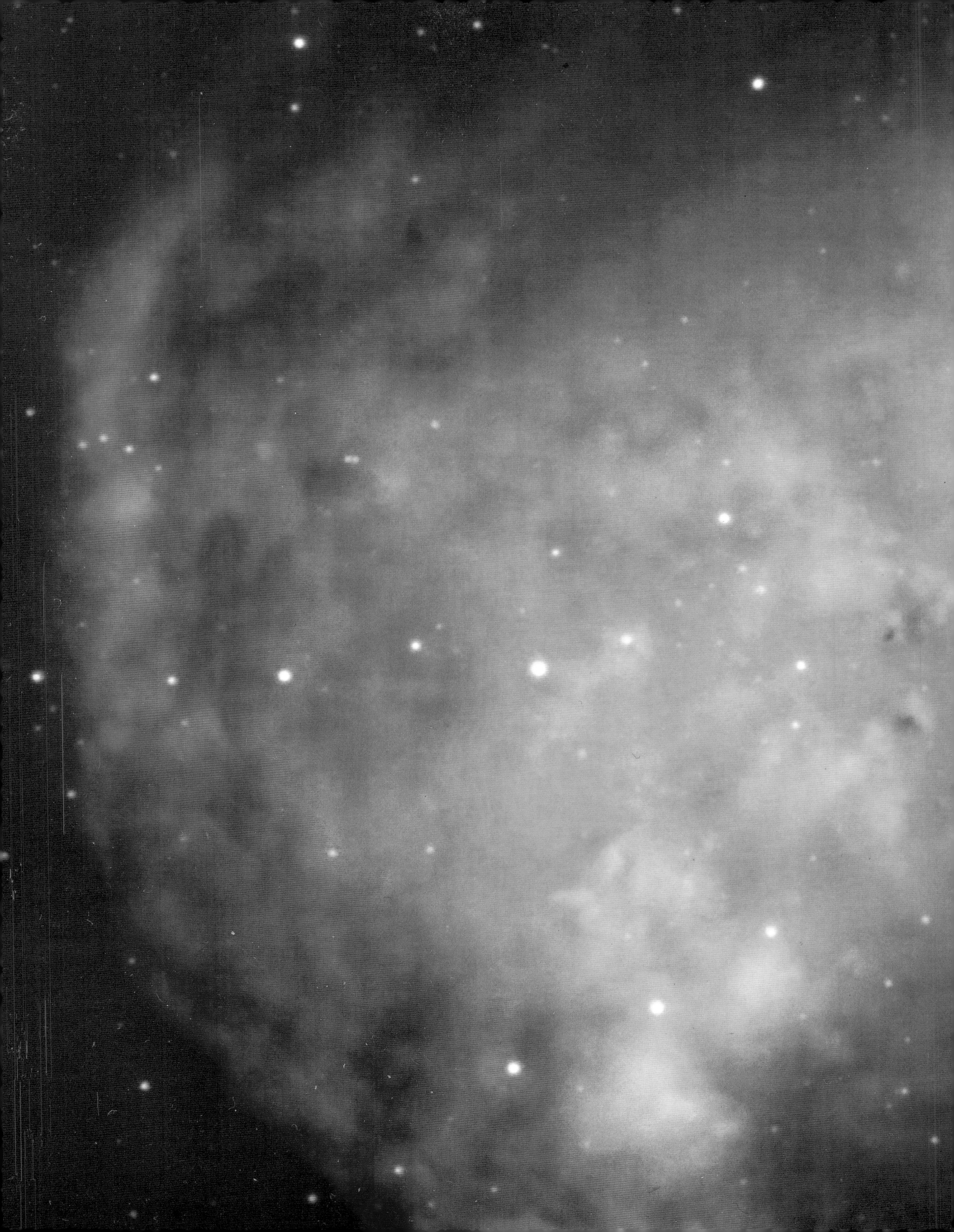